Synthesis Lectures on Human-Centered Informatics

Series Editor

John M. Carroll, College of Information Sciences and Technology, Penn State University, University Park, PA, USA

This series publishes short books on Human-Centered Informatics (HCI), at the intersection of the cultural, the social, the cognitive, and the aesthetic with computing and information technology. Lectures encompass a huge range of issues, theories, technologies, designs, tools, environments, and human experiences in knowledge, work, recreation, and leisure activity, teaching and learning, etc. The series publishes state-of-the-art syntheses, case studies, and tutorials in key areas. It shares the focus of leading international conferences in HCI.

Ron Wakkary · Doenja Oogjes

The Importance of Speculation in Design Research

Ron Wakkary
Simon Fraser University
Vancouver, BC, Canada

Doenja Oogjes
Eindhoven University of Technology
Eindhoven, The Netherlands

ISSN 1946-7680   ISSN 1946-7699 (electronic)
Synthesis Lectures on Human-Centered Informatics
ISBN 978-3-031-67094-7   ISBN 978-3-031-67095-4 (eBook)
https://doi.org/10.1007/978-3-031-67095-4

This Springer imprint is published by the registered company Springer Nature Switzerland AG
The registered company address is: Gewerbestrasse 11, 6330 Cham, Switzerland

If disposing of this product, please recycle the paper.

Acknowledgements

We would like to thank Audrey Desjardins, Jordan Espheter, Brett Halperin, Will Odom, Netta Ofer, Marie Louise Søndergaard, Oscar Tomico as well as the anonymous reviewers for their generous and thoughtful feedback on earlier drafts of the book. We also thank the designers, artists and researchers whose works we discuss for informing and inspiring our thinking, writing, and sensemaking throughout the book. Finally, we are grateful to our colleagues at the Everyday Design Studio, Simon Fraser University and the Making-With cluster at Eindhoven University of Technology for continuing to recognize and take seriously the importance of speculation.

Contents

List of Figures

Why Speculate?

Speculation is a form of reasoning that is underexplored in design research despite the adoption of practices like speculative design, critical design, and design fiction. Speculative design has been valued for examining future trajectories of designing with technologies. However, it has also fostered a perception that speculation is only about critiquing and futuring. It is further perceived as a genre of design rather than central to these practices. This distances speculation from the material and empirical investigations that govern design research, especially in fields like human-computer interaction (HCI). Even more consequentially, this limited understanding hinders the possibility to extend speculative reasoning to create new and liberatory approaches that address urgent challenges facing design researchers today.

In this book, we argue that speculation can be central to design research. We see it as working alongside or being interwoven with traditional forms of reasoning. Where traditional reasoning is problem-solving based on empirical evidence, speculative reasoning is based on the creative use of propositional knowledge. Traditional reasoning infers from defined or determinate evidence to support hypotheses or qualitatively determine findings. Speculative reasoning creates virtue in ill-defined or indeterminate evidence or by promoting multiple and simultaneous ways of knowing to reveal new insights and alternatives to traditional worldviews. Lastly, traditional reasoning in research is valued for its immediate utility or contribution to solving a problem whereas speculative reasoning is valued for making clear ethical consequences and its potential to politically challenge and to see the world anew. What we do not do in this book is argue that one form of reasoning is better than the other, we see them as simply different and argue that we need them both, and at times the line that separates them is unclear. Together, they offer a more expansive understanding of reasoning in design research.

© The Author(s), under exclusive license to Springer Nature Switzerland AG 2025   1
R. Wakkary and D. Oogjes, *The Importance of Speculation in Design Research*, Synthesis
Lectures on Human-Centered Informatics, https://doi.org/10.1007/978-3-031-67095-4_1

What do we mean by *design research*? The term is broadly used in different practices with similar but different meanings. This underscores the degree to which designing has become a part of many fields. Specifically, the term began in architecture and industrial design during the design methods movement (see (Jones 1970)), referring to research on improving design practices. Our use is grounded in its evolution and expansion to include a more mutual relationship between design and research in a shared and mutually informing practice. A practice investigates the creation, consequences, or applications of designed artifacts, systems, services, and technologies. Design research in this sense is practiced in a range of domains including computing, cognition, embodiment, socio-cultural systems, ethics, and the environment to name a few. Our understanding of design research is informed by our practices in the field of human-computer interaction (HCI) that maps well to our description of design research. HCI is commonly understood as the study of how better to design interactions between people and computing. Our use of the term may be grounded in HCI but we both look and practice outward toward and past the typical boundaries of HCI to include proximal and related design practices and ideas. We chose the term *design research* to signal the specifics of practices like HCI within and including the broader terrain of practices related to design.

Despite the value of speculation as we see it, many researchers are suspicious of the usefulness of theorizing without solid evidence or empirical data. In this light, speculation in design is seen to be at worst mere opinion or at best theoretical but in either case without practical utility (Forlizzi et al. 2018) or unable to contribute to knowledge in its own right (Bardzell and Bardzell 2013) or as a privileged practice (Burton and Nitta 2011; luizaprado 2017; Ansari 2019b). As a result, speculative reasoning in design is underdeveloped as a resource and seen as less important. This is not meant to minimize the important role that the genre of speculative design has played in mobilizing designers and researchers. Rather, we want to extend these gains by arguing that speculation reaches beyond the confines of a category called "speculative design." So how to argue that speculation is an important resource in design? To make the case that speculative reasoning is a central mode of thinking and designing alongside other modes of reasoning that together drive empirical and creative efforts? And that speculation is increasingly necessary in design to address current and future challenges such as the climate crisis, decolonization, and racism.

To begin with, to speculate is not to fantasize or act without reason. In the face of incomplete knowledge, speculative reasoning opens the knowledge space further and finds ways to navigate the ambiguity and contradictions of what is not known and does so with attentiveness and focus to contribute toward a better understanding of the world. Speculation creates possibilities that can be further reasoned on or acted upon.

In mathematics, speculation is part of what is formally known as a *conjecture*. A conjecture is a prior conclusion or a proposition that is suspected to be true but has yet to be proven or disproven by a mathematical proof. Pierre de Fermat in the 1600s offered what became known as *Fermat's Conjecture*. The conjecture excited and spurred new theories of algebra through to the nineteenth century and modular forms in the twentieth century. Eventually, some 350 years later, a proof was found. *Thought experiments* in physics or philosophy are similarly speculative. A well-known thought experiment is Albert Einstein's *train-and-platform* speculation on the relativity of time or simultaneity. Einstein imagines how two different perceptions of the same phenomenon of time can be true. Two lightning bolts hit the front and rear of a moving train car, and two observers, one observer in the moving train car and another standing on the train station platform, perceive the sequence of the lightning strikes differently. For one, the strikes occur simultaneously and for the other, the lightning at the back of the train strikes first, yet both are held by Einstein to be faithful observers. The experiment imagines two incompatibles but equally true perceptions of the same event.

The likelihood of one lightning strike hitting a train car is remote, but two lightning strikes! Imagination in speculation is not just filling in the blanks of what we could know but taking advantage of what we do know by creatively experimenting with it. The tenth century Muslim philosopher, Ibn Sina, known in the West as Avicenna, describes a thought experiment in his *Fi'-Nafs/De Anima* (Treatise on the Soul) known as the "flying man." In it, we are asked to suppose that we are created at once with no imperfections, but we cannot see anything external to us. We are a body flying through the air but we cannot feel the air, nor do our limbs and extremities touch each other. Despite this complete lack of sensation, we know we exist. Ibn Sina offers that such a flying man cannot affirm the existence of their body yet can affirm their existence of self. Some six centuries prior to Descartes's idea of *cogito* or knowing subject, the flying man thought experiment speculates the existence of the self-aware subject or knowing self.

1.1 Leaps of Imagination

Speculation is creative, it involves leaps of imagination and the commitment to pursue the possibilities that are made available through these leaps. Creative acts like mathematical proofs and thought experiments use imagination to look past what you can see and expect to see, to creatively add what you did not see or did not expect to see. Not one lightning strike but two. Not sensing one's body in flight but sensing one's self. The philosopher Isabelle Stengers, for whom speculation is important, writes how leaps of imagination are at the heart of the speculative philosophy of Alfred North Whitehead (Stengers 2011):

> To think with Whitehead is also to affirm that the success of a philosophical proposition is not to resist objections but to give rise to what he himself calls a "leap of the imagination" and the point is to experiment with the effects of that leap: what it does to thought, what it obliges one to do, what it renders important, and what it makes remain silent. (Stengers 2011, 22)

Speculative philosophy shows that the value of speculation is not that it is a "truth-in-waiting" to be empirically validated. Rather as a proposition, it creates an opening in which the effects of the proposition, its logical consequences, its diverse possibilities, potential common good, or felt experiences can be experimented upon and foster action. Einstein's theory of relativity waited over a hundred years for empirical validation. Fermat's Conjecture demonstrated an inordinate degree of patience, as it took over 350 years for a proof to be discovered. Yet each speculation led to new ways of knowing in physics and math during that time. Speculation is not mere problem-solving. While an answer may be forthcoming, the real power of speculation is that it creates the conditions to think with the unknown and act without answers. It creates a powerful space of experimentation and one that is arguably more impactful as a proposition than a solution. Further, as an unanswered proposition, it is open to suspend dominant ways of thinking that can lead to new knowledge and actions.

Staying with the philosopher Whitehead, Stengers and her colleague Didier Debaise offer a further nuance to what we mean by speculative leaps of imagination (Debaise and Stengers 2017). Whitehead's speculative thinking encourages us to "eliminate nothing." This echoes the importance of imagination to actively consider possibilities no matter how odd or illogical they may at first appear. However, this is not encouragement to engage imagination unhindered. Rather, Debaise and Stengers argue "Whitehead's speculative thinking aims at maximizing friction with experience…" (Debaise and Stengers 2017, 16). For them and Whitehead, to speculate is to leap into unseen possibilities of our world—to imagine with one's eyes wide open. Einstein's relativity thought experiment takes place in a physical world of moving trains and a stationary train station. Even Ibn Sina thinks through the corporeal body even though it has no sensations and can fly. Friction is the world of the probable, the world of dominant modes of existing. Within speculation, a leap of imagination is the aspiration and ability to see the possible in the probable without eliminating it but resisting it: "What we need to activate today is a thinking that commits to a possible, by means of resisting the probable…" (Debaise and Stengers 2017, 18). The unanswered proposition in speculation is brought about through leaps of imagination (proofs, thought experiments, scenarios, etc.) that creatively reveal what you cannot see or expect to see and critically work through the friction of the probable to see what's possible. In this way, leaps of imagination can suspend, while keeping in view, dominant ways of thinking to lead to new knowledge and actions.

1.2   Diverse Epistemologies

This latter power to propose alternatives to dominant thinking is more than a matter of creative strategies of flying bodies and lightning bolts but runs deeper to include diverse ways of knowing based on different assumptions and precepts. Speculation can be epistemologically diverse through seeking different ways to know the world. Robin Wall

Kimmerer, a botanist and citizen of the Potawatomi Nation, reflects on her first days as a university student studying botany:

> In moving from a childhood in the woods to the university I had unknowingly shifted between worldviews, from a natural history of experience, in which I knew plants as teachers and companions to whom I was linked with mutual responsibility, into the realm of science. The questions scientists raised were not "Who are you?" but "What is it?" No one asked plants, "What can you tell us?" The primary question was "How does it work?" (Kimmerer 2015, 41–42)

In her own telling, it would take some time before she was confident enough to hold her indigenous way of knowing—"a natural history of experience"—with equal if not greater weight to the dominant Western European ways of science. Kimmerer's book *Braiding Sweetgrass* (Kimmerer 2015) weaves together diverse ways of knowing that better articulate a deep and necessary ecological consciousness of the world we cohabit. In her book, there is a passage in which Kimmerer offers the traditional story of Nanabozho, the last being created or first human for Turtle Island, as told by the Anishinaabe elder Eddie Benton-Banai. Nanabozho as Original Man was instructed to walk the earth that was created and to name everything he encountered. In Kimmer's retelling of the retelling, she herself wanders to imagine Nanabozho walking together with the Swedish botanist Linnaeus, who formalized the binomial nomenclature of naming organisms firstly by genus and secondly by species. In Kimmerer's mind, they formed an excited and happy pair walking together. Linnaeus explained how in his taxonomy every organism is related and Nanabozho enthusiastically agreed saying that at one time all of Creation could speak together and knew each other's names. Linnaeus replied that common language has been lost and so he used Latin to name everything. Kimmerer concludes her speculative variation by saying that Linneaus offered Nanabozho his magnifying glass to see the exquisite floral details. In return, Nanabozho offered Linnaeus a song to hear the spirit of the plants (Kimmerer 2015, 209).

Kimmerer's wandering away from the traditional story becomes a speculation about traveling between worldviews or different ways of knowing. The speculation is in keeping both worldviews operative despite their contradictions and tensions. That is not to give in to some final verdict in which one is held up as knowledgeable and the other naive. Her speculative ability simultaneously draws on each way of knowing for their distinct explanatory and descriptive powers. And to not shy away from abandoning dominant ways of knowing the world for those that offer greater propositional force. It is on this deeper level, the willingness to accept that we always think and act with incomplete knowledge or know the world incompletely whether through western science or indigenous ways of knowing. And further, to see the worldviews not as mutually exclusive but speculating so we can fluidly travel between epistemologies.

Speculation is then performative; the willingness and ability to diversify our epistemologies to expansively reason with the world. The reward is to see the world differently, maybe for how it really is, by addressing it imaginatively. In her book *The Mushroom at the End of the World*, Anna Tsing tells the story of the Matsutake mushroom as a multispecies ethnography (Tsing 2015). Key to Tsing's more-than-human ethnographies is what she calls the *art of noticing*. For Tsing, the world is not occasionally punctuated by crises or moments of precarity, rather it is in a constant state of precarity with little promise of stability. This ongoing series of disturbances is brought about through the ruins produced by capitalism, whether its economic vulnerabilities, dispossessed peoples, or the increasing and constant climate crisis in which we are enmeshed. However, Tsing's speculations are not a dystopic critique, but rather, through noticing, she guides our attention toward the new forms of life and living of multispecies relations that arise within the ruins. She turns our gaze away from exclusively human solutions, idealized futures, or full faith in progress to attune our abilities to *notice* what is newly produced, namely, multispecies relations that form the basis for any future survival on this planet, though they are at the margins, largely unnoticed.

This is not an abstract proposition but a material speculation in which Tsing follows the Matsutake mushroom, a Japanese delicacy, across the disturbed and deforested landscapes of China, Japan, Finland, USA, and Canada. She traces the collaborative survival of fungi and pine trees that rely on each other for nutrients and protection. The pairing thrives in the human disturbances of deforested lands that remove cover to let pines draw in the unencumbered sun and rain, Matsutakes to lie protected at the end of the short roots of the pines, and economies of migrant human foragers to emerge that seek out the pairing. For Tsing, "telling stories of the landscape requires getting to know the inhabitants of the landscape, human and nonhuman" (Tsing 2015, 159) or to speculate on what is typically unseen or goes unnoticed. "It is time to pay attention to mushroom picking. Not that this will save us-but it might open our imaginations" (Tsing 2015, 19).

Making use of the power of speculation to widen our ways of knowing to creatively counter or add to what we know can also be seen to be a powerful space to help negotiate what is the good of that knowledge and for whom. Where Tsing calls our attention to new thriving relations, albeit in a world of precarity, speculation can also reveal new relations and ways of knowing even in the inequities and power imbalances of global economies or colonialist legacies.

The anthropologist Hirokazu Miyazaki's ethnographies of the Suvavou people of Fiji speculate on a method of hope that ties the generational efforts to gain full reparation of lands with the ideas of hope by the Marxist philosopher Ernst Bloch (Miyazaki 2004). The Suvavou people for well over a hundred years, across generations, repeatedly challenged colonial and postcolonial governments over the wrongful transfer of ownership of ancestral lands. At the center of the challenge is belief in the existence of a document to validate the claim on the lands that was either misplaced or hidden. Over the years, as an exercise in hope, the Suvavou petitioned colonial officers directly and in postcolonial times were left to repeatedly request access to and search the government archives. The archives became a postcolonial system of both oppression and liberation. The very proposition that such a document was there to be found, generation after generation, turned each repeated moment or present into a prospective moment in which the future could be different as a result. For Miyazaki, this repetition with the hope of a different future each time resonated with Bloch's argument that repetition is a purposeful method of hope in which new knowledge could arise. Further, while hope is directed forward into the future it requires a strong link to a past, and one that is also not fixed but reconsidered and learned again with each new search of the archives. In this way, Miyazaki sees in this performance of hope a process of self-knowledge, a reconstituting of past, present, and future within postcolonial legacies of power, structures, and resistance.

1.3   Ethical Reflexivity

A central relevance of speculation to design is that design practices are not only creative, but they can also be self-reflexive. We have so far discussed how speculation is the *leaps of imagination* that are inherent to creativity; and it is the *epistemological diversity* that supports and mobilizes self-reflexivity. This leads to the remaining reason for the importance of speculation to design that we focus on. Speculation can be a critical or self-regulating function on the limits of its own creativity. In other words, not all leaps of imagination are morally justifiable, e.g., nuclear weaponry. Epistemological diversity on its own is not inherently good or without potential for harm or injustices, e.g., phrenology and social policy. Such questions of the limits of speculation and the overall practices are questions of ethics and ones where the value judgment is not so apparent (as with nuclear annihilation or explicit racism of phrenology). Extending the self-reflexive capacity of speculation, we can see it as supporting *ethical reflexivity*. That is, to utilize speculation to investigate the limits or the consequences of our technical practices but to also build on these critiques to cultivate and develop an understanding of the *good* or pluralities of *good* in design.

In 1667, the French physician, Jean-Baptiste Denys performed the first documented human blood transfusion or more accurately, xenotransfusion, as it was the transfer of blood between a human and a lamb (Fastag et al. 2013). Prior to this, Denys had been experimenting with blood transfusions between dogs and between other animals including cows and horses. He was in a race to realize the possibilities of blood transfusions against English counterparts who were also experimenting with animal transfusions. Even more so, Denys, seeing himself as an outsider to the Parisian medical establishment, sought to show them up for their conservative and traditionalist skepticism of blood transfusions, in part caused by their own failed transfusion experiments. After having claimed success with his transfusions between nonhuman animals, Denys transfused blood from a lamb into the veins of a teenage boy who suffered from interminable fevers and extensive

Fig. 1.1 Illustration of Jean
Baptiste-Denys performing the
first documented
xenotransfusion of blood
between a lamb and human.
Source Public Domain

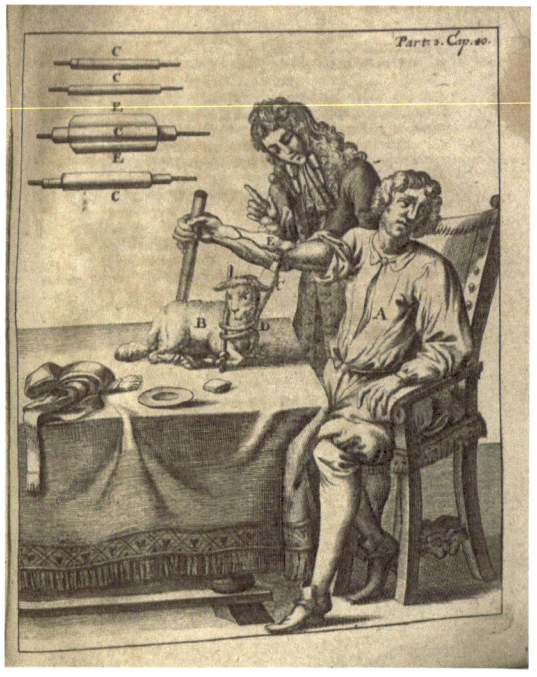

bloodletting (see Fig. 1.1). The boy survived and recovered. It is thought that minimal blood was transfused though indeed some had.

While other xenotransfusions likely had occurred, Denys's transfusion was the first documented account. It was in part recorded since he was brought to trial for the death of another patient who had unwittingly undergone several transfusions with calves' blood to remedy his perceived insanity. As the account goes, the criminal prosecution was partially funded by the medical establishment to discredit Denys, yet through the course of the hearings, the likely cause of death was determined to be poisoning by arsenic and not the blood transfusion. And the poisoning and murder were carried out by the patient's wife (Tucker 2012). As a result, Denys was acquitted of murder though the wife was less fortunate in this regard. Despite that victory, the judge ruled that human blood transfusions could only be administered by physicians of the Parisian Faculty of Medicine, of which Denys did not belong. In this strangest of stories, justice, politics and money were the odd bedfellows in victory. The degree to which this contributed to greater knowledge on the part of medical science is debatable since some two hundred years later, lamb blood transfusions would make a brief but wide-scale resurgence across Europe and USA. It ceased with the discovery of different human blood types and the adverse, if not fatal, consequences of mixing types together, let alone across species.

We tell this story to acknowledge the inherent and unpredictable dangers of the speculative leaps of imagination and unfettered experimentation. Yet it also reveals the unpredictable frictions and turbulence that can accompany speculation. In hindsight, exploits like Denys' are seen as quackery and rife with ethical lapses and lack of procedural checks and balances. Yet, "progress" has historically been very messy, littered with questionable ethical choices and risk-taking as a form of knowing. In medicine, over time, explicit and rigorous ethical and harm mitigation strategies have evolved. Though in this strangest of tales, medical research is not immune to the politics, power, and questions of justice and with outcomes we may not have expected.

1.4   Speculation in Design

Our discussion so far has been wide-ranging, giving free rein to the ideas of speculation. This was done to flip the script, so to speak. That is, to frame the space of speculation as open as possible to see design within the broader landscape of speculation rather than trying to understand speculation within the narrower confines of design.

Speculation already has a strong foothold within design in what is commonly known as *speculative design*. Speculative design raises questions that help negotiate the ethical implications of technology-related practices and the underlying question of the good of the practices. Related to the story of Denys' xenotransfusion, the designer Revital Cohen's *Life Support* (Cohen and Van Balen 2008) project includes scenarios that investigate the trajectory and possible limits of service relations with domesticated animals, particularly in areas of human health. Cohen asks, how far will we go in blurring the lines between companionship with and utility of animals, could we extend this to the point of companion species becoming medical equipment?

In one scenario, a lamb is a dialysis machine for a patient suffering from kidney failure (Cohen and Van Balen 2008). During the day, the lamb is free to roam the property, graze in the yard, and maintain its kidneys through drinking water and urination. In the evening, the lamb lies next to the patient connected together to circulate and clean the blood of the patient (see Fig. 1.2). As part of the fiction of the scenario, the lamb is transgenic as it was possible to create recombinant DNA in which the genome sequence for the blood and blood type of the patient could be infused into a sheep egg cell which is then carried to birth by an ewe.

Cohen's *Life Support* draws us into a fictional near-future that is rooted in the present-day trajectories of medical technologies, animal testing, and domestication. In doing so, the scenarios expose the values of the practices that led to this scenario, the latent possibilities within the probable. However, the ethical issues are offered ambiguously, for example, they draw comparisons between today's genetically modified foods or "free range" chickens for example. The speculation is a set of questions within a proposition: Given what is accepted today is this future acceptable? If not, why do we accept today's practices and their values? Do today's values lead to these outcomes? The propositional

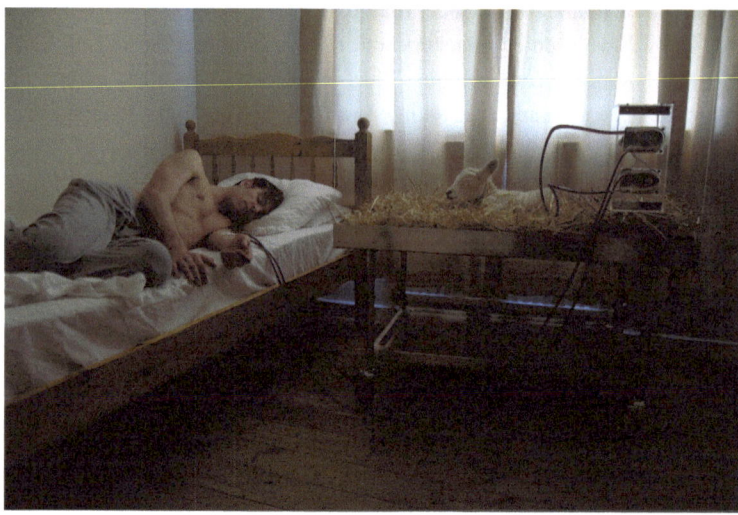

Fig. 1.2 Revital Cohen's scenario of a transgenic lamb dialysis machine in Life Support (2008). *Source* Courtesy of Revital Cohen

nature of the speculation maintains a reflexivity that is active and performative. The question of what is the good of the practice is framed as an imaginative opening that creates conditions for reflection.

This performative aspect is in keeping with the idea that the notion of "good" of the practices of design is not universal but rather negotiated and situated (Verbeek 2009; Puig de la Bellacasa 2017). This helps account for the changes of practices and technologies. More fundamentally, it acknowledges differences, or more specifically the inequities in what it is to be human—as what is good for the majority is all too often at the expense of those in the minority. The imagination of the proposition also expands the boundaries of ethical concerns. Cohen's *Life Support* directly positions ethical concerns of technologies beyond humans to include animals, as all too often, human values come at the expense of other species that cohabit our world.

As is clear with *Life Support*, speculation contributes to design research practices through leaps of imagination, epistemological diversity, ethical reflexivity, and experiential alternatives. The most recognizable of these would be leaps of imagination. Many readers would agree that design practices are creative. And agree that taking an imaginative leap is well established in practice when formulating hypotheses, illustrating scenarios of use, creating prototypes to test, or theorizing new concepts and applications. The genre of "speculative design" radically extends the use of scenarios and prototypes to investigate future possibilities. A hallmark of speculative design is the creation of experiential alternatives that investigate the boundaries of future possibilities, problematizing them rather than resolving them, and in the process making space for the other speculative processes of diverse knowing and ethical questioning. However, for all the success of speculative design (of which there has been plenty) and what it has to offer to this discussion (of which again is plenty and we will elaborate on in the next chapter), we find it important to separate *speculation* from *speculative design*. In this book, we commit to the idea that speculation is a form of reasoning that is applicable to all design research rather than a style or methodology applicable to only certain types of design research.

The importance of speculation in design research is firstly that it brings attention to the inherently creative aspect of knowledge production in design. That is the role that imagination, in particular, an expansive and diverse imagination, plays in creating *leaps of imagination*, or the productive spaces in which experimentation occurs and knowledge arises. Creativity can be argued as productively thinking and doing without full knowledge of the outcomes. Criticality is to both see and see *through* the frictions of the day to understand what is possible. Design in its variant forms brings creativity and criticality together, which is why prototyping, iteration, and testing are central concepts. More than not, leaps of imagination are made as prototypes and *then* put to the test. And this is true of not only the concrete technological artifacts and systems that are created but also the theories and concepts that drive the practices (Carroll and Kellogg 1989; Redström 2017).

Secondly, these very theories and concepts that enable design are known and accepted to be in a state of constant change. The practices of designing with technologies are epistemologically emergent, is one way to state it, and is seen as self-reflexive practices that constantly look outside of its dominant worldview or seek *epistemological diversity*. Some may argue that there is a need to stabilize theories in design research into a more singular foundation of knowledge (Oulasvirta and Hornbæk 2016). However, what is abundantly clear is that past efforts to theorize the discipline are full of shifts and changes. In HCI, the field is often described as a series of "waves" from first to second to third reformulations of accepted knowledge into something new, less known and less theorized (Bødker 2006; Harrison, Tatar, and Sengers 2007). Another frequently used term is "turn" in which design shifts its direction, pivoting toward new theories and priorities whether its experience (McCarthy and Wright 2004), materiality (Robles and Wiberg 2010), practice (Kuutti and Bannon 2014), or feminism (Bardzell 2010). Elsewhere, I (the first author) have argued through critical posthumanist theories, that design be seen less as a discipline and more pluralistically as a set of inexhaustible "nomadic practices" that are concurrent, distinct though overlapping (Wakkary 2021). For us, underlying these waves, turns, and critical unpacking of the discipline is a drive to epistemological diversity, a willingness to speculate on how design research can fluidly occupy diverse ways of knowing.

And lastly, we need to acknowledge and heed the harms that can arise from designing though in contrast to the medical harms of xenotransfusion, the cause of the harm in designing may be much harder to determine and the timing of effects latent though no less impactful. For example, the harms of social media are cascading effects that play out in increasing probabilities at scale whether its life-threatening mental illness or ethnic genocide (Hunter 2021; Mozur 2018). The danger of unfettered creativity and technological innovation is plain to see. Mitigation of such harm is no doubt implemented through societal change, government regulations, and legislation yet there is the need for ethical reflexivity within the practices of design itself (Wakkary 2021). A reflexive attention is necessary to guard against potential harm but also the construction or shaping of what is argued to be the good of the practices.

1.4.1 Speculation Framework for Design Research

To summarize, we have offered three ways in which design can draw from speculation. These include:

- Speculation mobilizes creativity in design through **leaps of imagination** that critically engage the friction of the world to experiment with the effects, to foster new ways of thinking, commitments, and matters of importance;
- Speculation fosters exploration and use of **diverse epistemologies** for design to be reflexive; to be critical of the dominant knowledge of the practices and open to the potential of alternatives;
- Speculation creates space for **ethical reflexivity** within the practices of design by analyzing and assessing the limits and consequences of the practices.

Speculation, its propositional power, serves a different purpose than validation or transferable knowledge. Speculation's purpose is to be propositional, to set the conditions for action, alternatives, reflexivity, or new directions achieved through creativity and imagination. The propositional nature of speculation is conceptually generative as in adding a second lightning strike or a flying man to a dilemma or investigation. In design, the additive function is material and is done through the creation of what we call, an *experiential alternative*, the equivalent to a conjecture or thought experiment. Each of the important qualities of speculation: leaps of imagination, diverse epistemologies, and ethical reflexivity are methodologically enabled through the making of an alternative experience:

- Speculation is materialized as **experiential alternatives** that methodologically enable leaps of imagination, epistemological diversity, and ethical reflexivity.

Such experiential alternatives can be material, conceptual, or performative, in essence, any specific realization of a speculation. The result is to set conditions for possible action and thought rather than resolving situations. An experiential alternative offers a material case to be negotiated, debated, that can function beyond exclusively textual investigations. For example, Pinar Yoldas' *Ecosystem of Excess* is a speculative biology or what she refers to as "anticipatory zoology" (see Fig. 1.3). In the work, she creates new organisms that she

Fig. 1.3 Pinar Yoldas'
ecosystem of excess (2014).
Source Photo by Ron Wakkary

imagines emerging from the "primordial soup" of plastics that float in our oceans. The speculation is that "our waste" has an ongoing life and intervenes in other life forms in ways not only toxic but unknown. The material realization of the imagined forms makes visceral the proposition—the possibility of what occurs through our actions and beyond our limits.

1.5 Our Approach to Speculation

Our approach runs counter to most discussions about speculation and design. Most scholars subsume speculation within the genre of speculative design which includes critical design and design fiction. The result is a limiting of the role of speculation. For example, discussions focus on the boundary limits of speculation particular to a sub-discipline of design (Malpass 2017; Tharp and Tharp 2019) or a design methodology (Koskinen et al. 2011; Bardzell and Bardzell 2013). Implying that other sub-disciplines and methods do not draw on speculation. Others aim to define the goals and outcomes of speculative practices to make clear what is and what is not within the reach of speculation. For example, speculative design is future-oriented, fictional, critical, and not art (Dunne and Raby 2013) or as some argue, speculative design is exclusively focused on science and technology futures (Malpass 2017); whereas critical design as a methodology is required to propose holistic change, be grounded in theory, improve public competence and be aware of itself as a social actor seeking change (Bardzell and Bardzell 2013) or attend to present-day social, cultural, and ethical implications of design practices (Malpass 2017); the purpose of design fiction is to make storytelling prototypes and build worlds to critique possible futures (Sterling 2009; Dunne and Raby 2013; Lindley and Coulton 2015; Bleecker 2009); and lastly, a variant like discursive design is about audience reflection (Tharp and Tharp 2019). We ask why can't speculation be about all the above and more? Rather than expanding our understanding of the potential and reach of speculation, these approaches narrow our understanding of speculation. By delimiting speculation into sub-disciplines or mutually exclusive genres and outcomes, a taxonomical problem of overlapping and contradictory definitions arises. This compels scholars to sort out and fix the different boundaries like parceling out land or categorizing species (Malpass 2017; Bardzell and Bardzell 2013; Dunne and Raby 2013). In contrast, we turn to expanding the features, goals, and outcomes of speculation in design. Viewing speculation as an investigative process or activity rather than a sub-discipline or genre and one that is central to reasoning such that it is potentially important to any research in design.

1.5.1 The Aims of the Book

In this introduction, we make the case that speculation has a central role in research. By speculation, we mean creating propositions about the world with incomplete knowledge. The value of speculation is that it can help construct propositional knowledge and actions within the limits of what is known at the time. Speculation holds even more value to counter accepted knowledge or dominant worldviews to move beyond blind spots. Speculation can seek out less privileged alternatives or new forms of knowledge. And equally important, it creates space to examine and assess the limits, consequences, and ethics of the pursuits of technological practices.

Beyond acknowledging the role of speculation in research, we aim to show how design research is building on speculation and can continue to build on it in ways unique to its practices to address emerging and new matters of concern. And so, the overall aims of the book are to further develop the role of speculation in design, better understand how speculation has functioned in past and current practices, and argue how speculation can be called upon to function in future practices. In this way, we make the case that speculation is not just important now but will be increasingly important in facing urgent new challenges.

As we have tried to make clear in this introduction, our aim is to approach speculation as expansively as possible and in doing so, explore how design can draw even more from the imaginative processes of speculation. With respect to the specific practices of speculative design, we aim to learn from the myriad historical examples and evolved techniques that materialize many aspects of speculation. To use these as a solid starting point to build on by drawing from other design research that fall outside of what is commonly referred to as speculative design but offer wider claims to epistemological diversity, reflexivity, and ethics. This includes features of speculating on not only the future but also the present and the past. We see speculation within the domains of science and technology but also topics such as philosophy, climate change, racism, and decolonization. Rather than speculation being a methodology it spawns a wide range of methods including futuring, fiction, critical methods, critical fabulations, speculative enactments, material speculation, alternative presents, co-speculation, and so on. And we expect researchers to continue to experiment with and expand methods based on speculation.

The overarching goals of the book are to uncover the potential importance—and centrality—of speculation to design research such that it helps practitioners and researchers better understand their practices and where they may go. A second aim is to argue for an increased importance of speculation in these future practices. Both present and future practices that need critical imagination, reflexivity, ethical imagination, and the means to negotiate what is important and to whom.

1.5.2 Outline of the Book

The book is organized in three chapters plus this introduction and the conclusion. The three chapters are the heart of our argument. As a rhetorical structure, each is organized by temporal framings of the future, present, and past. This allows us to begin with what speculation is most associated with in design research: speculating on the future. But then to explore how speculation also has a key role in investigating present-day conditions of technology and technology use to seek latent alternatives or strategic interventions. And last to focus on speculations of the past, to make pasts provisional so as to unpack how dominant assumptions, power relations, and technological trajectories came into place or shaped the present and future. This speculating on the past opens possibilities of alternate grounds from which to consider technology design or to reveal hidden realities that have shaped current-day practices in ways unseen. These temporal framings metaphorically show the reach and expansiveness of speculation in current and past practices but also potential new ones.

Relatedly, we appreciate how language positions time in relation to when a phrase involving time is uttered, e.g., the future is only in relation to the moment a phrase is spoken or written. This acknowledges not only the relative nature but also the relational nature of time. This is in keeping with speculation, as it underscores that to speculate on the future equally speaks to the possibilities of the present, and that speculations of the past can alter our understanding of present-day realities, and so on. With this in mind, we chose to use the *continuous* verb tenses of English grammar (though also found in many other languages) to metaphorically capture our use of the future, present, and past.

Further, each chapter is organized by our framework for speculation discussed earlier. This structure explores the range by which speculation utilizes leaps of imagination, explores epistemological diversity, creates ethical reflexivity, and experiential alternatives.

We will explore key methodological approaches used to create experiential alternatives. Lastly, aim to demonstrate how aspects of the speculations we highlight can be traced to existing design research.

Chapter 2: *Continuous Future* shows how speculation can be a future orientation that is creative and critical. We begin this chapter with the Italian Radical Design movement that formed the crucible from which practices of speculative design, critical design, and design fiction emerged. Here we illustrate the critical role that speculation can play in research. For example, the leaps of imagination of Italian Radical Design or Afrofuturism in design, explicitly resists probable formulations to reveal obscured possibilities. We discuss how speculation in these approaches as well as others in speculative and critical design, engage future trajectories of the status quo world to question these futures or offer alternatives. We highlight a key method for experiential alternatives we call *para-functionality* that refers to functionality figuratively or metaphorically. A recurring theme is how speculation resists dominant and probable futures. We extend this to see how resistance and skepticism toward dominant thinking can be traced to design research that is commonly not seen as speculative. For example, Mark Weiser's vision of ubicomp can be seen as a critique and propositional alternative for the future of technology research at the time. More prosaically, we discuss how critical literature reviews or even participatory design methods speculatively resist or critique dominant future trajectories.

Chapter 3: *Continuous Present* reveals how speculation has a role in material and empirical investigations in design research. It offers the understanding that what is experienced and materialized may hold unresolved complexities or can even obscure conditions of the present. We discuss how a range of design research uses strategies such as alternative presents, material speculation, and speculative enactments to examine, question, and engage present-day understandings of issues and technologies. Speculation in this chapter is not anticipatory of futures but rather reveal conditions of the present, enable critical encounters, and permit friction. The chapter highlights a key method for experiential alternatives we call *things that work otherwise*, artifacts and system that function in atypical circumstances falling a non-dominant rationale. We conclude by extending this understanding of speculation to a range of approaches in design research from STS readings of design practice and feminist perspectives to hypothesis testing and ethnographically informed breaching experiments.

Chapter 4: *Continuous Past* shifts attention to the past. Not to give a historical account but to make the past provisional, open to speculative re-interpretations. We describe how these speculations open the past to reveal hidden realities that shape current-day realities or offer new possibilities for the present day. Our approach in this chapter is to see how speculation can be a liberatory mode of inquiry for future design research. We focus on speculative practices of feminism, Afrofuturism, more-than-human epistemologies, disability studies, and decolonization as they relate to design and technologies. We highlight a method for experiential alternatives we call materialized stories that reenact the past imaginatively and critically. In short, in this chapter, we balance between analytical takes on past research to more of a prospective, even speculative outlook, of what design research could address.

Lastly, we conclude with Chapter 5 in which we summarize the arguments we made for the importance of speculation and how it expands reasoning in design research. We elaborate strategies and methods to help put speculative reasoning into practice. We discuss the implications of this on ongoing and future research practices, and how researchers can further build on the work we have presented.

Continuous Future

2

The opening line of the poem *Darkness*, written by Lord Byron, reads "I had a dream, which was not at all a dream. The bright sun was extinguish'd." The poem describes a sunless sky, a world in famine, and the end of humanity, as told by the last man, the narrator of *Darkness*. Byron's poem speculates about human extinction, when "waves were dead, the tides were in their grave," the moon had "expir'd" and the "winds were wither'd."

The poem was written in 1816 during the "Year Without a S'mmer." A time of severe and abnormal climatic change that created darkened skies and freezing temperatures across the planet, including Switzerland, where Byron wrote the poem. He spent that summer in a chalet on Lake Geneva with the gothic novelist Mary Wollstonecraft Shelley and her husband, the romantic poet Percy Shelley. Unknown to them, the shrouded skies were caused by a volcanic eruption of Mount Tambora on the island of Sumbawa, some 10,000 km away, in what Europeans at the time called the Dutch East Indies and is now Indonesia. The eruption occurred a year earlier, spewing over 100 km^3 of ash and material that over months drifted through the earth's atmosphere to obscure the sun and plunge much of the planet into darkness. The eruption devastated the islands of Indonesia and caused massive famines throughout Asia. Crop failures occurred in Europe. The severity and scale of climate change caused panic and a sense of doom. For Lord Byron and the Shelleys, there was the realization that humanity could end, a feeling of such finality that had largely been unimagined until that point.

In response to the possible end of humanity—its extinction—the writers set out to create expressions of *horror* that could potentially capture what was beyond their immediate grasp. Byron wrote the poem *Darkness* to put to paper what up until then had not been expressed—the horrors of human extinction. Wollstonecraft Shelley began writing

R. Wakkary and D. Oogjes, *The Importance of Speculation in Design Research*, Synthesis Lectures on Human-Centered Informatics, https://doi.org/10.1007/978-3-031-67095-4_2

Frankenstein; or, The Modern Prometheus that was to be published in 1818, the well-known cautionary tale of a young scientist giving life to a hybrid human-machine. While a horror story, its techno-scientific underpinnings make it one of the first examples of science fiction (Aldiss 1995).

For speculative practices, *Darkness* and *Frankenstein* are antecedents of leaps of imagination through fiction and explorations of unknowable futures. The poem and science fiction/horror novel together picture not only an unimagined future but one of unintended or unimaginable consequences—a future to resist. *Darkness* grapples with colossal climatic change-again a future to resist. *Frankenstein* especially, draws out the anxieties, chaos, and human peril latent in the space between the creation of technologies and the impact of those technologies. And most importantly, the form by which the imagination takes hold is a speculative account of a near future expressed in horror, like a warning or caution of a path taken or a choice before us. And so, not unlike a chain of genetic mutations, the DNA of the gothic novel mutated into science fiction to branch into a set of speculative practices in design that share a position of resistance when looking to the future.

Since futures are inherently less graspable than the present or past, it makes intuitive sense to rely on speculation to create propositions about possible worlds. Given this, speculation is often seen as synonymous with conceptualizing possible futures, a creative way to offer up provisional futures as propositions for analysis and reflection.[1] Of course, as discussed earlier, we aim to show that speculation is applicable to inquiries of the present and the past as well. Nevertheless, it's valuable to begin this book with understanding how speculation in design helps attend to future concerns. It offers a common sense understanding of speculation that can readily be extended to design research, fields that are generally preoccupied with researching technological artifacts or systems that don't yet exist. However, good research needs to be critical and skeptical about what might be designed or might exist.

In this chapter, we will show how speculation affords design a future orientation that is both creative *and* critical. The examples we will discuss carved a path for what is often referred to as the "genre" of speculative design. We bring these together as Continuous Futures: speculation oriented toward the future, continuous and relational with the past and present, though resistant to dominant futures. However, as discussed earlier, we aim to move speculation beyond genre or sub-discipline status. Our discussion begins with the Italian Radical Design, which formed the crucible from which practices of speculative design, critical design, and design fiction emerged. Speculative design helps to illustrate

---

[1] With speculation, we find an interesting relationship between the provisional and propositional. At times, speculation seeks to make known thoughts provisional, to destabilize certainties to open us up to propositions of what it could be like otherwise whether speaking of the present, past, or future or the relations between. This relationship is especially clear in Chap. 4 in which speculation is used to unseat accepted historical narratives, making pasts provisional to create propositional ideas of the present and future.

the inherently critical nature that speculation can play in research. Drawing on what we earlier described as the leaps of imagination, speculative design explicitly resists probable formulations to reveal obscured possibilities. It engages the friction of the status quo world and its trajectories for the future to offer alternatives or open these futures up for questioning. Here speculation is put to work to resist the hegemonic forces and dominant modes of thought to open for analysis, what should be designed and what should not. A leitmotif throughout this chapter on continuous futures will be how speculation *resists* futures, and how from a wider angle we can see how resistance and skepticism toward status quo or dominant thinking is common to design research, well beyond what is commonly seen as speculative.

We will use the framework of speculation explained in the introductory chapter to examine how this future orientation of speculation, the continuous future, is enacted as *leaps of imagination, diverse epistemologies*, and *ethical reflexivity*. We draw out the characteristics of *experiential alternatives* that materialize in these practices and the methods behind them.

2.1 Leaps of Resistance: Futuring and Fiction

In 1966, 150 years after the natural disaster of the "Year Without a Summer", the Arno river, some 600 km to the southwest of Lake Geneva, overflowed and flooded the city of Florence. It was the worst flooding of the Arno since the 1500s. After long periods of rain, on November 4, 1966, fearing that an upstream dam would collapse, some 2000 m^3 of water were discharged to travel at some 60 km an hour down the river into the historic city center. At its highest point, the mud and water reached almost 4 m above street level. The designer, Adolfo Natalini was forced to abandon his studio in the city center to join his friend and fellow designer, Cristiano Toraldo di Francia, at the outskirts of Florence. It was during this time that the pair founded the design collective known as *Superstudio* (Didero et al. 2017). The collective was the first of many in a movement that came to be known as *Italian Radical Design*. The movement would spawn similar collectives like *UFO, Archizoom, Gruppo Strum,* and *Gruppo 9999*.

The late 1960s was a politically turbulent era globally with the US led Vietnam War in Southeast Asia, the genocidal murders of the communist and ethnic purges in Indonesia, the Tlatelolco massacre in Mexico City, the Bloody Sunday of the march from Selma to Montgomery, the anti-war protests in Berkeley, the October crisis of the Quebec separatist movement, and the student and worker uprisings in France and Italy. Philosopher Rosi Braidotti characterized the movements as the rise of anti-humanist thinking, challenging the inheritances of humanist or enlightenment thinking that she saw as underpinning the legacies of fascism, Soviet communism, and late-stage capitalism—movements that sought new epistemologies, social theories, and a more radical politics (Braidotti 2013, 17). The radicalism of the period cannot be overstated, nor the chaos and uncertainty that accompanied it. In Italy, the level of political violence and social conflict that emerged was unrivaled in Europe. The years that followed the student and worker uprisings became known as the *Anni di piombo* or "years of lead," given the endless string of assassinations and horrific bombings culminating in the 1978 kidnapping and murder of former prime minister Aldo Moro by the left-wing extremist group the Red Brigade (Statera 1979). It was during this era that the Italian Radical Design movement was most active.

In Italy, like much of Europe, student enrollment in universities doubled in the 1960s setting the stage for student activism (Statera 1979). The students questioned the dominant capitalism of post-war industrialism—in spite of or because of the rapid economic growth and industrialization it brought to Italy—that shifted capital into private hands and exploited both labor and landscape. Across Italy, student occupations and strikes at university campuses agitated for educational reforms and later wider democratic reforms; aligning with workers, students moved to occupy factories as well. The movement was strident and utopian, though this would eventually give way to ideologically driven terrorism by both neo-fascists and militant left-wing groups. Nevertheless, the anti-authoritarianism and anti-establishment of the movement formed a deep-rooted resistance to mainstream society that would galvanize a new generation's political consciousness and desire for wholesale political change.

The architecture department of the University of Florence was a center of student activism. Natalini and Toraldo di Franca were students in the department, along with others who would become part of the Radical Design movement. Soon after their decampment to the outskirts of Florence, and after the flood waters had subsided, the newly formed *Superstudio* combined with the collective *Archizoom* to create an exhibition titled "Superarchitettura" in nearby Pistoai. Translating the newly formed political consciousness into design led to a merciless anti-design position. The manufacturing and design industry in Italy were on the rise, globally exporting a style of Italian modernism. Successful companies like the office product manufacturer Olivetti or the car manufacturer Fiat, were targeted as agents of mass consumer capitalism that were complicit with and masked the increasing cultural and military imperialism of the West that the students sought to resist.

In response, the Superarchitettura exhibition disavowed the functionalism and underlying consumerism of modernist design. The exhibition included only images of objects painted on cardboard or plywood that had no function and only a vague purpose. Toraldo di Franca aimed to unmask the complicity of design: "the perverse activity of the system in continuously reproducing poverty, new desires, and waste" (Hucal 2016). This internal provocation against design from within design would make a refrain some thirty years later in Anthony Dunne and Fiona Raby's opposition of critical design to affirmative design (Dunne and Raby 2001), in which the affirmative design is status quo and consumerist. A manifesto graced the poster for the exhibition. It was an absurd negation of design that dripped with irony:

the superarchitecture is the architecture of superproduction, of superconsumption, of superinduction to superconsumption, of the supermarket, of superman and super-petrol. (1966)

The leaps of imagination of the Italian Radical Design were leaps of resistance within a political and historical moment that enacted resistance in the present—in the streets—with aspirations of a radically different future. Resistance materialized socially in political acts of disturbance and occupation and materialized in designing as non-functionality, impermeability, and ultimately anti-design. Imagination at the service of resistance became emblematic of what followed Italian Radical Design through to today in the varying and overlapping practices of critical design, design fiction and speculative design. Here, resistance takes the form of speculation into an enterprise of dismantling the sociotechnical imaginaries (Taylor 2004; Jasanoff and Kim 2015) of the day to reveal the obscured possibilities of knowing and action. Speculative resistance served the goals of utopian possibilities, empowering those at the margins, and liberation through hopes of dismantling the hegemonic forces.

2.2   Diverse Epistemologies of Resistance: Critique and Aspire

Other Florentine students including Lapo Binazzi formed the group *UFO*. The collective pursued design as a subversive program that included repeated interventions into public spaces they called *the Urboeffemeri* series (Didero et al. 2017). These guerilla actions or "happenings" occurred throughout 1968 often alongside student demonstrations. The *urboeffemeri* or "urban ephemera" involved taking to the streets of Florence with objects made of poor materials like oversized inflatables meant to mock and challenge the surrounding stone monuments and architecture (Hucal 2016). In the fifth of these happenings, a large inflatable in the shape of a rocket was emblazoned with the phrase "Colgate con Viet Cong" (see Fig. 2.1), marrying the advertising slogans of a toothpaste company with the sloganeering and weaponry of the Vietnam War. In an interview, Binazzi expressed how the inflatable was the result of a free association or wordplay that was a "crossing" of the languages of political news and advertising. Epistemologically, the absurdity resisted the differences of the languages to dismantle a corporate slogan to reveal the instrumentalizing of war for corporate ends. For Binazzi, the interventions were a call for "imagination as a political experience" (van Balen 2018).

For many of the Italian Radical Design adherents, architecture and product design represented the design establishment that needed to be undone. This emerged as a radical critique and utopia that did away with buildings, products, and design as a whole. This is the negation of status quo knowledge in architecture and design–to build buildings and create products. In a 1971 lecture at London's Architectural Association, Natalini expressed Superstudio's disavowal of architecture and design:

Fig. 2.1 Urboeffimero nr. 5 by
U.F.O. (1968). *Source*
Courtesy of Lapo Binazzi

> If design is merely an inducement to consume, then we must reject design; if architecture is
> merely the codifying of bourgeois model of ownership and society, then we must reject archi-
> tecture; if architecture and town planning is merely the formalization of present unjust social
> divisions, then we must reject town planning and its cities . . . until all design activities are
> aimed towards meeting primary needs. Until then, design must disappear. We can live without
> architecture. . . . (Lang and Menking 2003, 167)

Superstudio disassembled architecture into anything but a building. A way of asserting
architecture and design as a site of resistance to the status quo to allow for new forms of
knowledge to emerge. A resistance that could come in the form of a mocking critique like
UFO's inflatables that targeted the self-aggrandizing, rationalist, and totalizing towers of
modernist architecture.

This mocking critique was often accompanied by a fantastical utopianism that can
be seen as both skeptical and aspirational. For example, Superstudio created a series of
photo-collages of natural and urban landscapes that are wholly or partially covered in an
infinite, totalizing architectural grid system titled *Supersurface: An Alternative Model for
Life on Earth* (see Fig. 2.2). The grid can be seen as a satire of the unadorned streamlined
and modular stylings of modernist architecture—unabashedly dominating the landscape
as a totalizing force. Though within the critique, is a utopian vision of a world devoid of
consumer objects and production, free of human labor, open to the nomadic wandering
of inhabitants in which people "live with objects and not for objects" (Frassinelli et al.
1972, 245).

This *resistance to design through design* is a radical departure into new territory that is
critical of and challenges design's argued replication and support of hegemonic powers.
The opening of epistemological possibilities through speculation is achieved through irony
and critique: political disruption and mockery; and partly through aspiration: fantastical
utopia. All as the means to dismantle existing assumptions and in their place to create
an *opening* for alternative epistemologies rather than *asserting* a particular alternative

Fig. 2.2 Supersurface: an alternative model for life on Earth by Superstudio (1972). *Source* Courtesy of Art Resource

epistemology. Again, by epistemology, we mean the ways of knowing and the measures by which knowledge claims are seen to be credible and valuable. These speculations not only question what is *good design* but also what is the *good* of design, meaning what are the moral and political contributions of design. This resistance is enacted through speculation in that it aims to construct a question to demarcate space for potential answers without giving the answer or solution. The utopias envisioned seemed so fantastical as to put them into doubt or surround the solutions in scare quotes. While the particularities of the approach may be different, this shares the same power of mathematical proofs and thought experiments to mobilize a space of possibilities, albeit in a socio-political space.

The mock irony of the Italian Radical designers would be refined in later speculative design into practices such as critical design (Dunne and Raby 2001; Dunne 2008) and design fictions (Bleecker 2009). The radical critiques and fantastical utopias of the Italian Radicals would give way to more nuanced critiques and explorations of future at the edges of possibility (Dunne and Raby 2013). Nevertheless, they are equally leaps of imagination into unseen political alternatives as critique and resistance to ask the question of such possible futures. Not unlike Wollestonecraft Shelley and Byron, it is a creative questioning in response to an uncertain horror that for the Italian Radicals was the political establishment, and for speculative design is often the technological world and its underlying political, social, and economic arrangements.

Yet radical scrutiny of the day, including political and utopian aspirations, never left the continuous future of speculation. This reflects the ongoing aim of revealing and mobilizing obscured epistemologies and ideas of different futures. Kirsten Bray and Christina Harrington co-designed a speculative design toolkit with Black and LatinX youth known as the "Building Utopia" design cards (Bray and Harrington 2021). The cards bring together speculation and participation to "reimagine potential futures" by decentering dominant systems of power. They created tools that ground speculative futures in the "lived experiences of Black and brown communities" (Bray and Harrington 2021, 1793), and so squarely center race by asking speculators to consider the absence of environmental racism. The deck contains cards enabling speculators to define their own liberation, situated within concerns of Black communities such as "black joy, health equity, and community safety," and cards that focus on equitable futures that address injustices of the past (see Fig. 2.3). The explicit grounding in Afrofuturism (Bray and Harrington 2021; Bray et al. 2022) and Critical Race Theory (Ogbonnaya-Ogburu et al. 2020) is a diversification of epistemologies, centering marginalized views of the world made transparent and offered as the conditions of speculation. Key among this approach is the notion of "Counter-storytelling" that makes clear that narratives of the future should resist and separate from dominant narratives and assumptions (see Fig. 2.4).

Whether it is horror, resistance, or political oppression, these speculative leaps traverse the tightrope of speculative thinking not by offering specific solutions or analysis but rather choosing to open imaginative spaces for alternatives. Imagination is rooted in politics aiming to decenter power for those at the margins. Speculation can mock, agitate, and empower to aspire to a utopian future without knowing what that utopia might be but ensuring its grounded in addressing concerns and injustices of the present and past (akin to Miyazaki's *method of hope* of the Suvavou discussed in the previous chapter (Miyazaki

Fig. 2.3 Building Utopia cards consists of five types of cards plus blank cards: (1) liberation, (2) topics, (3) technology & solution, (4) design methods, (5) blank cards, and (6) forecasting. *Source* Courtesy of Kirsten Bray and Christina Harrington

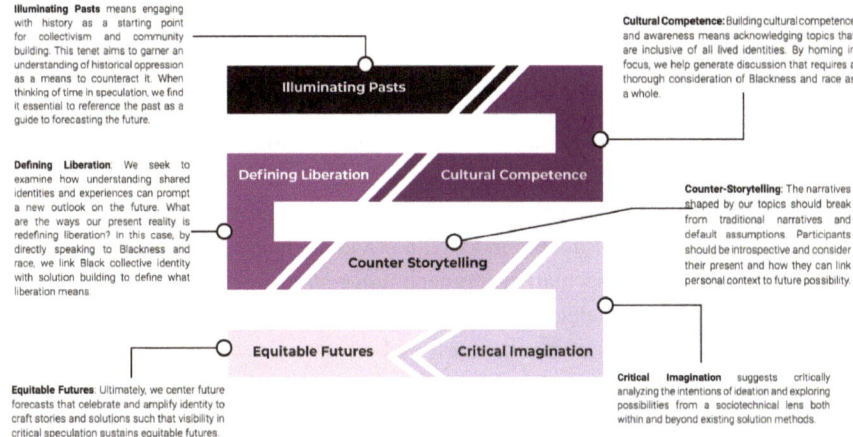

Fig. 2.4 Bray and Harrington's proposed approach to integrating tenets of critical race theory in Afrofuturist speculation in design. *Source* Courtesy of Kirsten Bray and Christina Harrington

2004)); speculation in design embraces irony, critique, and imagination to investigate possible technological futures, avoiding solutions and prescriptions, opting to view criticality like resistance as speculative, a process of creating multiple possible alternatives and diverse epistemologies.

2.3   Provocations of Ethical Reflexivity

The leaps of imagination we discussed created openings in which resistance is intended to open alternative sites of utopian possibilities, foreclose dystopian trajectories, or more than not somewhere in between, sites of critical awareness in need of alternatives. As discussed, these sites are intended as imagination as politics. However, the anti-functional and ironic strategies emblematic of speculative resistance are not intended to produce concrete actions or solutions. While a lack of concrete action and solutions can be seen to undermine the political effectiveness of speculation in design (Tonkinwise 2014; Ansari 2019b), the ambiguity of what is offered is an intrinsic feature of speculative work. As we discussed in the opening chapter, the power of speculation is that it creates the conditions to think with the unknown and act without answers. It can be a productive space, an unanswered proposition that suspends dominant ways of thinking that can lead to new knowledge and actions. In Italian Radical Design, speculative or critical design and design fiction, this openness to interpretation is often materialized in objects (Sengers and Gaver 2006; Gaver et al. 2003), whether real, fictional, functional, or not. With our next examples, we unpack how this openness enables ethical reflexivity.

An early example of what came to be known as speculative design, *Placebo Project* (Dunne and Raby 2001) by Anthony Dunne and Fiona Raby, exemplifies the ambiguity of speculative objects and the ethical reflexivity this openness can foster. The project includes eight objects designed to investigate people's experiences of electromagnetic radiation. The objects include the *electro-draught excluder* that is a board made of medium-density fiberboard (MDF), a handle, and a grid of cone-shaped foam that ostensibly deflects (or protects) one from electromagnetic fields. Another object is the *GPS* table that if not visible to satellites, typical of a table indoors, a screen embedded in the table surface displays the word "lost." The speculative approach relies on the ambiguity of the objects that emphasize the invisibility of electromagnetic waves to reflect on the anxieties, instrumental logic, and potential harms of the then-new and increasing use of wireless technologies. *Placebo Project* also shows the diffusion of resistance from the immediate, visceral, and situated politics of the streets of Florence in Italian Radical Design to more generalized sites of resistance of the social and political forces that underpin technologies such as wireless and mobile technologies.

This broader speculative resistance to what is often called sociotechnical imaginaries is to speculate on the complex interdependencies that shape the use and determination of social norms through technologies. The speculative resistance is crystallized into objects (or imaginaries) for ethical reflexivity. By sociotechnical imaginaries, we mean the coshaping of collective social values and design of technological systems, as well as the normative commitments of given societies that are reflected in the technological project (Jasanoff and Kim 2015). A good example of a sociotechnical system is the Internet. Pierce and DiSalvo (2017) see in the metaphors we use for the Internet like net, web, superhighway, global village, or cloud, different "collective visions, hopes, and dreams concerning what the Internet is and might become" (Pierce and DiSalvo 2017, 1384). They engage these metaphors through speculative design to seek alternative metaphors to understand, interrogate and reimagine the Internet (see Fig. 2.5). The metaphors as objects of ethical reflection aim to amplify and channel "undesirable, uncertain, and ambivalent feelings surrounding digital networks" (Pierce and DiSalvo 2017, 1384). For example, the "Internet is Eye" signals our obsessive need to look at our devices as well as the unceasing surveillance and tracking of our actions through the Internet; or the "Internet as Hole" could refer to the endless alternative realities we find ourselves in while using the technology yet conversely also signals the need to escape to a material reality and find refuge from the technology. Some metaphors point to desirable futures, while others point to futures to resist and avoid.

The *Menstruation Machine* (2010) by Sputniko! (Ozaki 2010), also known as Hiromi Ozaki, portrays a wearable device that allows anyone, whether physiologically menstruating or not, to experience menstruation. The speculation is presented as a story in the form of a pop music video (see Fig. 2.6). In the video, Takashi, who wants to experience menstruation, builds the menstruation machine. The machine, worn like a chastity belt, can dispense blood and through electrodes stimulate the lower abdomen to cramp in pain.

Fig. 2.5 Alternative metaphors from Dark Clouds, Io$#!+, and 🔮 [Crystal Ball Emoji]. *Source* Courtesy of James Pierce

After completing the machine, Takashi gets ready to go out for the evening wearing the machine. In the video, they prepare themselves with a shiny orange wig, lipstick, purse and other adornments of the Japanese *kawaii* aesthetic or cuteness. The electro-pop music of the video fits the aesthetic and throughout the video, the lyrics of the song interject with questions about the desire to menstruate and Takashi's experiences of pain and blood during the evening in amongst questions from a girlfriend that is also out on the town with Takashi. The storyworld of the menstruation machine in the form of a pop music video is convincingly shaped through the aesthetic details of the music, colored neon lighting of the arcades they visit, the kawaii fashion, and taking of selfies in a photo sticker booth or *purikura*.

The *Menstruation Machine* creates a device that propels the story to become the object of ethical reflexivity by resisting normative assumptions of gender and technologies. The intent of the work is not to create a functional machine to let the wearers experience menstruation; that is only the storyline to get us to ask questions. The provocation is the propositional nature of the speculative resistance that creates alternatives and commits to exploring them. Here, even the designer is asking questions alongside the audience. The *Menstruation Machine* is to move beyond our technological world in which technology, gender, and biology converge as matters of social functioning like birth control and in ways that reify gender norms. The propositional space is purposely kept open like in a thought experiment and so multiple interpretations are inherently part of the speculation (e.g., see Bardzell et al. 2015). This propositional space for reflection includes a world

Fig. 2.6 Still from the video for the Menstruation Machine. Courtesy of Hiromi Marissa Ozaki

in which bodily technologies are designed for desire more than function, regardless how incongruous or unusual it is to some that menstruation could be desirable; a world in which gender is non-binary, and technologies augment us physiologically to support gender as performance; and a world in which bodily technologies are DIY and as personalized as individual fashion and style.

2.4   Experiential Alternative of Para-Functional Things

As described in the introduction, experiential alternatives are particular to speculation in design research. For example, critical theory can also speculatively resist by using theoretical frameworks that reveal and dismantle status quo assumptions to make way for better alternatives. However, unlike design, the speculation remains in the realm of text and language and more than not is analytical and descriptive. Design by contrast draws on its capacity to materialize and give form to concepts and ideas in ways that are visceral, multi-dimensional, and even messy. In continuous futures, this capacity crystalizes resistance in modalities other than speeches and texts by creating, for example, mocking inflatables and seethingly visual manifestos, products, and music videos that make the speculations and resistance visceral and experiential in ways distinct from language. Arguably, this experiential aspect, its realization in a material sense, can better explore the space of possibilities than language alone. Even ideas that are never fully realized materially (for example, depicted in photo collages or videos) are vivid trailers or previews of a fuller realization that seems possible, potentially functional and thus experiential.

We feel that the propositional nature of speculation makes it inherently generative as in adding a second lightning strike or a flying body to a dilemma or investigation. Experiential alternatives are a methodological approach unique to design. They are the design equivalent to a mathematical conjecture or a thought experiment. Each of the important qualities of speculation: leaps of imagination, diverse epistemologies, and ethical reflexivity are experientially enabled through the making of an alternative experience.

In the speculations of this chapter, that we have called continuous futures, we see a common approach to the making of experiential alternatives that we call *para-functional things*. This refers to the making of technological artifacts or technological referents that function figuratively or metaphorically to make an alternative world experiential. This experiential alternative expands the role of technologies beyond functional use to support speculations that resist norms and subvert everyday assumptions. The term "para-functionality" was first used in design by Dunne (2008). Para-functionality is a method that turns functionality in on itself by rejecting it outright, putting it to odd use, or by creating a representation of functionality without actually working. The method resists seeing design exclusively in terms of functionalism and utilitarianism to subversively expand the space of possibilities and imagine other worlds.

A disavowal of functionalism is part of Italian Radical Design's anti-design position that rejected a functionalism gone awry in the service of imperialism and capitalism. This is evident whether through the non-functional objects of the Superarchitettura exhibition or Superstudio's rejection of architecture as functional structures (Figs. 2.1 and 2.2). However, moving past the rejection of functionality to an ironic embrace of its qualities enables a broader reach for speculation. Dunne sees in functionality a rhetorical power and referred to para-functionality as an *estranged functionality* that through putting functionality to unconventional ends or purposes opens a rich and critical discussion of alternatives. For example, Dunne and Raby's GPS table implements navigational technologies in an indoor object that rarely moves, while Hiromi Ozaki details the making of a machine to simulate menstruation (Fig. 2.7). The GPS Table makes visible for discussion not only wireless technologies but the illusion of the reach of such technologies to track all things in the world though stymied by the simplest and most commonplace arrangement as a table in a home. As discussed, the *Menstruation Machine* reflects on the role of technology in health and gender.

The specific crafting and material shaping of para-functional things is critical to their success. The experiential alternatives are not conceptual placeholders but rather made "real" through material means like with the *Urboeffimero* by UFO or through the careful considerations and execution of technical and formgiving details in the *Menstruation Machine*. Para-functional things are quite liberal in interpreting what is an artifact or system, like with Superstudio's *Supersurface: An Alternative Model for Life on Earth*. Similarly, Pierce and Paulos' alternative metaphors for the Internet utilize emoji's as the main technological thing that while not exactly technologically functional, emojis have a very specific communicative function within different technologies. Pierce and Paulos carefully abide by the specificity of the design and aesthetics of emojis to undermine them in part to open to alternative meanings. The creative strategies and technical range of what constitutes a para-functional thing are also quite open, from visual design to interactive products. The Building Utopia cards by Bray and Harrington are outliers with respect to para-functionality yet they rely on the history and practices of Afrofuturism that utilizes para-functionality of technologies to envision liberatory ends and critiques of power and marginalized ownership of functional technologies.

Whether the things in continuous future speculations work or not is beside the point, rather it is the rhetorical power they offer to subvert and build alternative worlds. We will discuss in Chap. 5 the accompanying strategies and methods to these speculations including subversion and worldbuilding. For now, we turn to the broader field of design research to see how the characteristics of speculation discussed in this chapter are no strangers to many design research practices. With respect to experiential alternatives of para-functional things, there is a clear kinship with the role of prototypes that are so common to design research. Similarly, prototypes are often used to embody and explore possible futures in ways that functionality is not a prerequisite.

2.5   Continuous Futures in Design Research

It's easy to see how speculative design came to be seen as separate from traditional practices of design or research. While design research is prospective in nature, most design research is not explicitly about questioning the future. Speculative design filled a gap, asking questions of what *should* be designed rather than exclusively focusing on what *could* be designed. Or by asking *what else* can we know, rather than what do we know? However, these questions, as is speculation, are much more prevalent in design research than we might think. While the examples in this chapter are explicitly speculative, we see many examples in which speculation is more latent. We resist seeing the enterprises as distinct but rather interconnected.

Mark Weiser's Ubicomp is an obvious example of the use of speculation to create a propositional future for computation. Weiser's idea of ubiquitous or invisible computing resisted what he saw at the time as the trend to design the "dramatic machine," a computer so compelling we would never want to be without it (Weiser 1994). By contrast, ubiquitous computing decomposes itself into our environments, embedding itself so naturally into our world as to become invisible (see Fig. 2.7). The power of ubicomp as speculation cannot be understated. It is the Fermat's Conjecture or the train platform thought experiment of digital technology. It has been consequential in setting the dominant trajectory and sub-trajectories of technology development and use ever since. Its vision-centric approach underscores the future orientation of much design research, establishing methods like design scenarios (see Lunenfeld 2003) and prototypes (see Koskinen et al. 2011; Matthews 2014) as mainstay. Some critiqued ubicomp research for being in this constant near-future state overlooking the realities and the messiness of the present (Bell and Dourish 2007; Tolmie et al. 2002). And to further underscore the role of speculation in design research, Paul Dourish and Genevieve Bell argued that ubicomp could be read through

Fig. 2.7 Computer scientists at Xerox Parc with early prototypes to explore ubiquitous computing. Licensed under CC BY 4.0

science fiction in which researchers blurred technological facts with fictional futures; or as Julian Bleecker succinctly stated: in ubicomp, "fact follows fiction" (Bleecker 2009, 15). The ubicomp vision is very much a leap of imagination.

Another aspect of speculation, diverse epistemologies can be found in the most traditional of research techniques: a critical literature review. For example, in late 2009, DiSalvo et al. analyzed 157 papers related to sustainability in design research (DiSalvo et al. 2010), a relative explosion of research at the time, given that the notion of "sustainable HCI" was a recent emergence (Blevis 2007; Mankoff et al. 2007). The critical review utilized genre analysis to sort the literature into "genres" of shared commitments. The genres represent key underlying disciplinary foundations, such as social psychology in support of persuasion and behavior change; or engineering and computing science to inform technology design. The authors scrutinized the dominance of these intellectual commitments to speculate on missing perspectives such as rhetorical communication as a basis for behavior change or Science and Technology Studies (STS) to address the politics of technology design. Similarly, Desjardins et al. (2015) conducted a genre analysis of research literature on domestic technologies to create a speculative opening for future inquiries into the home such as non-anthropocentric or first-person perspectives (see Fig. 2.8). Seeking future alternate epistemological perspectives to resist or at least augment dominant trends is an ongoing pattern in design research, for example making a

The Protagonist Observer

WHY THESE TYPES OF PICTURES?

The observer aims to understand the experience of the home first hand.

WHERE IS THE OBSERVER?

The observer is positioned in front of the camera, or he points it towards his own experience of the home.

VISUAL CHARACTERISTICS

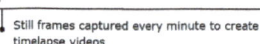

[19]

Photographs taken from the point of view of the researcher in his or her home

[12]

Still frames captured every minute to create timelapse videos

The Absent Observer

WHY THESE TYPES OF PICTURES?

The observer aims to capture the experience of the home from the perspective of non-humans (including objects or pets in the home).

WHERE IS THE OBSERVER?

The observer is completely removed from the situation and lets non-humans observe and report on the experience of the home. The researcher's role is in creating and installing those new methods enabling non-humans to gather data about the home.

VISUAL CHARACTERISTICS

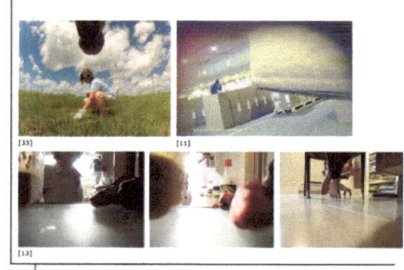

[33]   [11]

[13]

Photos of everyday scenes, from angles that are new to a human observer

Fig. 2.8 Two alternative perspectives to expand the researcher perspective for HCI research on the home by Audrey Desjardins and colleagues. *Source* Courtesy of Audrey Desjardins

case for postcolonial computing in response to design for the "developing" world (HCI4D) (Irani et al. 2010) or reassessing design for its colonial legacies to open to decolonizing design (Ansari 2019a; Nemer 2022), or modulating the ideals of *making* to include the realities of *unmaking* (Sabie et al. 2022).

Arguably, the continuous future of design research has a long history of resisting dominant trends and to provisionally open to new epistemological turns to dramatically shift the field as a whole. At its most ambitious is Terry Winograd and Fernando Flores' argument against the rationalist tradition of designing technologies generally, and AI specifically (Winograd and Flores 1987). They bring to bear a mix of philosophical traditions including hermeneutics (e.g., Martin Heidegger and Hans-Georg Gadamer; Heidegger 2008; Gadamer 1976), cybernetics (e.g., Maturana and Varela 1991), and language (e.g., Austin 1962) to undo the positivistic, rationalist basis for design in favor of an interpretivist and pragmatic view of the way technologies work. Philip Agre, similarly offers what he calls "critical technical practice" to shift the field's focus from classic cognitive science model of the interior mind to "interactionism," a mutual engagement and reciprocal shaping between humans and environment (Agre 1997). In design and HCI in particular, this critique of dominant trends and use of speculation to redirect the continuous future of the field manifests as "waves," from Bødker's identification of the "third wave of HCI" (Bødker 2006) to more recent claims of a posthuman wave in what Chris Frauenberger coins as "entanglement HCI" (Frauenberger 2019). The merit of these interventions has borne out over time as the field has expanded to include a diversity of epistemological perspectives.

Ethical reflexivity is by no means exclusive to speculation, often seen as a matter of research integrity that includes a values orientation. Yet, speculation makes ethical reflexivity an outcome of the process, it externalizes the ethics at hand as well as internalizing it within an integral research process. A good example of this beyond the genre of speculative design is critical approaches to participation in design, specifically participatory design or co-design. For example, John McCarthy and Peter Wright investigate the politics and aesthetics of participation in design (McCarthy and Wright 2015). In considering the value of participation they highlight the need for *dissensus*, a form of participation that challenges hierarchies and consensus. Drawing on the French philosopher Jacques Ranciere, they see dissensus as an "innovative leap" that overcomes what normally governs our affairs. This is not unlike leaps of imagination we discussed earlier; however, the emphasis for Wright and McCarthy is how through design the conditions for dissensus, a form of speculative resistance and ethical reflexivity, can be created in a participatory design sense. They cite their work with colleagues to co-design bespoke probes with participants to investigate personhood in the context of dementia (Wallace et al. 2013). Others have utilized speculation to create a space for reflection through participation in workshops (Andersen and Wakkary 2019) or imaginaries (Kim et al. 2022). Similar explorations are closely related to speculation in design such as utilizing agonism (DiSalvo 2012), discursive practices (Tharp and Tharp 2019), and "design for debate" (Mollon and Gentes 2014; Arets 2024) to create conditions for reflexivity. In many ways, Bray and Harrington's explicit combination of speculative design with participatory design (Bray and Harrington 2021) brings this trend full circle.

Techniques used in these traditions relate to the experiential alternatives of speculation, including scenarios and prototypes that are at the heart of ubiquitous computing. The range of rhetorical techniques like metaphors, genres, and fictional vignettes to mobilize theoretical interventions like the various waves of HCI. Embodied practices like workshops, debates, design cards, and what has come to be known as "imaginaries" are also related to the experiential alternatives of speculation in design research.

Our aim in this section is not to fold all design research into speculative reasoning. Seeing these practices in this light hopefully reveals how speculative and scientific reasoning have been interwoven together in many design approaches to research. There are, as we said, interconnections that hold commonalities as well as differences, but the characteristics of speculation are evident throughout design research practices.

2.6   Summary

In this chapter, we focused on speculation that is future-oriented. We characterized this aspect of speculation as resistant or critical towards prevalent, dominant, or probable future trajectories, that interweave politics, technologies, and cultures, by offering propositional futures to create or maintain space for alternatives. We described the experiential alternative of para-functional things as a common method for the designing and creating of artifacts and systems. We largely drew on what has come to be known as speculative design beginning with Italian Radical Design through to present-day work, though our aim is to view speculation not as a genre or sub-discipline of design. As we will see, this chapter is the first installment of our investigation of the meaning of speculation in design research. We applied our criteria of leaps of imagination, diverse epistemologies, ethical reflexivity, and experiential alternatives to draw out the particulars of speculation as well as the strategies used to create experiential alternatives. Arguing that the characteristics of speculation are important to broader research in design, we emphasized the interconnectedness of speculative aspects in a range of consequential to methodological arguments.

Continuous Present

3

In 1915, Virginia Woolf wrote: *"the future is dark, which is the best thing the future can be, I think."* This statement, though possible to read as daunting or dreary, is with its addition of *"I think"* also a declaration of committing to uncertainty, to engage with doubt and the unknown without the purpose of reaching resolution. Woolf was living in uncertain times in the midst of the First World War, battling through an episode of her recurring depression. In her essay "Woolf's Darkness: Embracing the Inexplicable," Rebecca Solnit unpacks Woolf's journal entry not as one of despair (which, Solnit argues, is another form of certainty) but as a declaration that asserts *"that the unknown need not be turned into the known through false divination"* (Solnit 2014, 86). Plans, knowledge, and authoritative language are only propositions of a reality that can be tried on for fit—and they are easy to make. Much more challenging is to engage with the obscure, the unknown, and the dark: *"and yet the night in which distinctions and definitions cannot be readily made is the same night in which love is made, in which things merge, change, become enchanted, aroused, impregnated, possessed, released, renewed"* (Solnit 2014, 86). Today, we find ourselves again living in uncertain times, with pandemics, wars, economic uncertainty, and climate change to name a few. And so, turning to the dark, the unseen and the unknown, and embracing its nuances and ambiguity can help us to understand the present isn't as resolved as it appears to be and reveal those relations that we take for granted.

> I hold in my hand a linden leaf, then an alder, willow, poplar and maple. No elms. I was promised elms. I look at my grandfather as he walks along the Admiralengracht before he met me, before he met my mother or my grandmother. I tell him: "that Elm won't be there much longer". He puts the leaf in his pocket and later, back at home—the address that is stamped on the cover page of the herbarium—he writes: young elms.

© The Author(s), under exclusive license to Springer Nature Switzerland AG 2025   37
R. Wakkary and D. Oogjes, *The Importance of Speculation in Design Research*, Synthesis Lectures on Human-Centered Informatics, https://doi.org/10.1007/978-3-031-67095-4_3

I hold my much smaller hand in his, my other in my grandmother's, as I glide down a slide. "I know now," he says, "Elm's disease." "Yes," I say, "but there are young Elms near our place now, at the Lauriergracht," and we walk over together.

Above's writing is a response to the second author's experiences in revisiting and re-enacting her grandfather's herbarium—a book of dried leaves from trees in Amsterdam he collected in 1933, when he was 17. We never had that conversation, not really, but in the past year, I have been engaging with the trees (present or not, living or dead), the city, and my grandfather by going on walks to find the trees he encountered and collecting material to create my own herbarium. The herbariums create an ambiguous space between the living and the dead that can be said to be speculative. It is a way of acting out the unknown, death (of trees or family members), as if it were materially knowable, and brings one closer to the deceased. In her work on relationships with the dead, philosopher Vinciane Despret turns to seances as an engagement that emphasizes the particular mobilization of the experiential (Despret 2021). Committing to the seance, by thinking about the dead and talking to them, is a way of creative thinking that surpasses the dichotomy of rationality and the extraordinary: *"artifice and authenticity of the experience are no longer contradictory modes. On the contrary, the better the artifice is cultivated, the more the experience will be lived as authentic"* (Despret 2021, 104). In the vignette above, there is a speculative leap in how the Elm trees are known through non-linear time. While the conversation never happened as such, the relationships with the Elm trees are based on the very situated and material act of creating herbariums in 1933 and in 2024.

This chapter aims to reveal the role of speculation in material and empirical investigations in design. What speculation offers in concert with empirical investigation is the understanding that what is seen, experienced and manifested is not always complete or stable. What might appear resolved—metaphorically dead—is able to be brought "back to life" into the present, whether through herbariums or seances to be engaged with. And so, the question becomes: how can speculation take face-value empirical matters, norms, and routines as clues or starting points for the obscured?

In this chapter, we focus on speculative practices, including Alternative Presents, breaching experiments, material speculations, and biographical prototypes. The speculations in this chapter are not anticipatory of futures but rather reveal conditions of the present, enable critical encounters, and permit friction. We will examine the workings of speculation again through the framework of *leaps of imagination, diverse epistemologies, ethical reflexivity,* and *experiential alternatives.*

3.1 Leaps of Imagination that Reveal Presents

In August of 2006, multiple pigeons flew over San Jose, California, wearing a back-pack equipped with sensors and communication units, to collect air quality data that was posted on a blog in real time. The pigeons were *co-producers* in the PigeonBlog project (see Fig. 3.1) led by artist-researcher Beatriz da Costa, a collaboration in which artists, engineers, pigeon fanciers (hobbyist pigeon breeders) and pigeons collectively engaged in collecting and distributing air quality data to make it accessible to the public. The mul-tispecies project positioned the pigeons as active collaborators, where much thought and planning went into learning to interact with, train and care for the pigeons—as well as into the design of the backpack with electronics with the weight and size requirements to keep it safe and comfortable. *"The pigeons became my communicative objects in this project and 'collaborators' in the co-production of knowledge,"* Da Costa writes in her artist statement (da Costa n.d.).

Donna Haraway sees the collaboration between pigeons, caretakers, researchers, and artists as an example of a practice she calls SF: science fiction, speculative fabulation, string figures, and so far (Haraway 2016). This practice engages with the idea of *staying with the trouble*, a leap of imagination in the way it conceptualizes the present as always becoming, the mundane and the messy as provisional. Staying with the trouble is a com-mitment to the situated and ongoing, a practice of *"becoming-with in times that remain at stake, in precarious times, in which the world is not finished and the sky has not fallen–yet"* (Haraway 2016, 55). This speculative leap is a commitment held centrally in the examples discussed in this chapter.

Fig. 3.1 Beatriz Da Costa's PigeonBlog project. *Source* Courtesy of the estate of Beatriz da Costa

The PigeonBlog project garnered much attention, with da Costa receiving requests for co-authoring a grant proposal for developing surveillance technology with birds and PETA attempting to shut the project down, seeing it as abuse without justifiable, scientifically grounded experiments. Pigeons have a long history in military use and law enforcement, and the PigeonBlog project itself was inspired by a wearable camera developed by a German engineer in the early 1900s for military purposes. However, PigeonBlog takes an artistic and activist stance in its reconceptualization. In response to PETA's argument grounded in the legitimacy of using animals for science, da Costa poses the following question: *"Is human-animal work as part of political action less legitimate than the same type of activity when framed under the umbrella of science?"* The PigeonBlog project is an example of amplifying and illustrating existing relations, while simultaneously revealing and inspiring new ways of collaborating. The leap of imagination in the PigeonBlog project is to take seriously and commit to the collaboration between pigeons, pigeon keepers, artists, engineers. This becomes clear in the reconfiguration and reconceptualization of wearable technology for pigeons, which took multiple design iterations and took into account the comfort of the pigeons. In looking for a more caring and co-constructive relationship between pigeons and people than the existing trajectory of weaponry and spyware, da Costa illustrates what designer James Auger might articulate as an Alternative Present (Auger 2010). Alternative Presents combines temporal technology lineages and speculation as a way to break free of assumptions that built incrementally over those trajectories, to instead bring to light and question those assumptions in the present. One of their well-known projects is a critique on the miniaturization and ubiquity of telecommunication, the *Audio Tooth Implant*: an implant that enables one to receive private audio messages, presented in the form of a mock-up scale model, live demonstration and media campaign. The leap of imagination in the project relies *"on a combination of general public awareness of hard and well-publicised facts such as the miniaturisation of digital technology, urban myths such as dental fillings acting as radio antenna and picking up audio signals and the rising popularity of mobile communication technology"* (Auger 2010, 8). What is revealed through Audio Tooth Implant are social and medical concerns and questions of desirability of implantable technologies, and current conditions of telephone communication and the implications of phones becoming smaller and mobile. For the PigeonBlog project, the Alternative Present of the collaboration revealed unquestioned or obscured use of animals in military technology. Both projects allow for critiquing otherwise unquestioned assumptions of technological development.

The leaps of imagination in these projects bring attention to unquestioned assumptions, and to do this, they rely on material encounters. Yiying Wu's Plant Hotels (Wu and Koskinen 2022) is a good example of such a material encounter, in this case through plants. Wu and Koskinen create what they call *Plant Hotels* that act as a service for travelers to maintain and care for their plants while away. Owners leave a plant in the hotel with a card (a Plant Story) that identifies the plant species and describes how to care for it (see Fig. 3.2). On the same card, a stranger can record what actions they took to care for the plant, add the date, their name and any additional notes. The hotels create a community of strangers to care for plants. A plant hotel is a leap of imagination that proposes an alternate form of social interaction and care among strangers. This break from routine interactions and care was inspired by the ethnomethodologist Harold Garfinkel and his *breaching experiment* method. In the method, researchers purposely disrupt the normal flow of day-to-day life to uncover what makes life routine (Garfinkel 1964). For example, Garfinkel encouraged his students to return home and act as a renter in the home rather than a family member creating a surprised social distance that allowed the researcher to study the response by the family. The plant hotels were opened in four locations including a university building, a conference event, a neighborhood gallery, a care home, and a fifth fictional hotel at the border between North and South Korea. Like a breaching experiment, the aim of the project is to uncover social structures, in this case, the formation of communities. In addition, Wu and Koskinen wanted to understand the role design could play. The plant hotels reveal perceptions and possibilities that otherwise go unnoticed and position these as something that can be designed toward. The fifth, fictional hotel (see Fig. 3.3) is a further imaginative leap but is treated with the same seriousness as the other hotels. The encounters with the plant hotels that exist in the actual and material realities allow the imagination to construct the reality of the fifth Plant Hotel. For example, when asked about the social interactions around the plant hotels, caretakers explained that they enjoyed looking after the plant but were less interested in the social aspect, in other words, the plants relied on social bonds but the actual people stayed in the background. The fifth hotel in Korea builds on this idea as an opportunity for citizens of both sides to come together and have peaceful conversations without direct contact. The plant hotels are *"prototypes of a potential new form of social action that they rendered visible and observable for research"* (Wu and Koskinen 2022, 36).

In an example of our own, The Tilting Bowl is a breaching experiment of sorts in probing expectations of what things in the home can be (see Fig. 3.4). The *Tilting Bowl* is a ceramic bowl that, as the name suggests, tilts only very slightly a few times a day. It is like any other bowl in that it can hold things like fruit, paper, keys or anything else you might put in a bowl. Yet unlike other bowls, the Tilting Bowl also houses a motor and offset wheel at the bottom that activates its tilting mechanism. The strangeness of having a thing so familiar as a bowl to have a technological component opens the door to speculate on the strangeness of other everyday objects having technology embedded, such as tweeting toasters, sensing fridges, and camera-enabled doors. Like Alternative Presents, it reveals the unquestioned assumptions about what is considered normal or strange in technological developments.

These modes of breaching, countering, shifting, and displacing create frictions that enable a propositional space, one that makes use of expansive imagination or speculation. The examples in this chapter make the current day provisional, not just the future. This cultivates openness and uncertainty by creating material experiences situated in the everyday that enable dialogues of alternatives and other propositions.

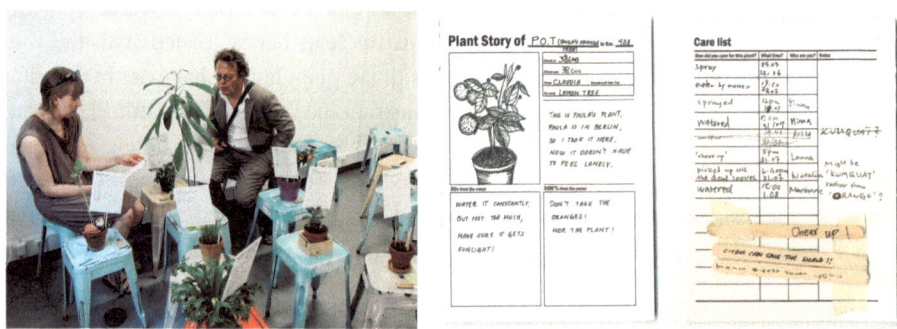

Fig. 3.2 Yiyung Wu and Ilpo Koskinen's plant hotels, including their plant story cards. *Source* Courtesy of Yiyung Wu

3.2   Diverse Epistemologies of Friction

The examples in this chapter bring together different ways of knowing the world in situated and material ways: the PigeonBlog takes scientific practices to open it up to the public, Plant Hotel combines the caring actions of strangers to re-imagine the forming of communities in a politically fraught location. The diversity in ways of knowing plays out in particular locations and field sites. Sarah Fox and co-authors' *RIOT project* (Reparative Internet of Things) is a study investigating the potential role of design and technology in maintaining infrastructures in the context of making menstrual hygiene resources publicly available. The project brings together the imaginative space of internet-connected everyday things (IoT) and labor studies. At the center of the project is a simple design (see Fig. 3.5) of a cardboard enclosure containing a networked sensor that counts the number

Fig. 3.3 The fictional fifth plant hotel as envisioned on the border of North and South Korea. *Source* Courtesy of Yiyung Wu

Fig. 3.4 The Tilting Bowl. *Source* Photo by Henry Lin

of menstrual products (such as tampons and pads) inside the dispensers and communicates this to maintenance staff as well as the public.

> What is and isn't considered vital to sustaining a healthy public? Who is expected to supply the material to support it? In some ways these questions were made more fascinating by the fact that we only discovered them at what might be considered the end of a traditional design encounter: the deployment. (Fox et al. 2018, 29)

Fig. 3.5 The RIOT (Reparative internet of things) being installed and calibrated for deployment. *Source* Courtesy of Sarah Fox

In the above quote, the authors reflect on the questions and tensions that arose when the researchers were setting up the deployment study of RIOT in the public university building. It was met with both skepticism and enthusiasm by maintenance staff and passersby. Maintenance staff recognized complaints about the often unavailable supply of menstrual products, yet also described the struggles, bureaucratic hurdles and labor the device could generate. The questions that arose during the situated inquiry emphasize the epistemological commitment to creating friction in everyday life, to situated inquiry. Here we see how Continuous Presents can reveal different levels of responsibility and labor that are assumed to support networked devices such as these. The productive friction is enabled by encountering these conditions in the world, known through the materialization and actual physical presence of the *RIOT* design. The maintenance worker's way of knowing was one of work, hurdles, and complaints. What is revealed through this project is the diverse epistemology of repair, feminism, and its connections to labor.

The very situatedness of the project plays a vital role here: it is important whose worldviews are engaged with. Biographical prototypes also aim to reveal invisible labor (Bennett et al. 2019). Cynthia Bennett and co-authors recall the story of kitchenware brand OXO founder Sam Farber, who created a vegetable peeler with a larger, easier to grasp handle. It is a well-known and often told design tale of how designing for one can be a way to design products that work better for everyone—in this case, Sam's wife, Betsy Farber who has arthritis. What is obscured in this telling of the tale is the labor and design work of Betsy, who articulated the need for a better handle and conceptualized the product. Cynthia Bennett responds to this by introducing the notion of biographical prototypes:

"*material demonstrations of how individuals with disabilities have made, adapted, and repurposed their environments and objects to work for them*" (Bennett et al. 2019, 44).

Through interviews and prototyping workshops, the researchers collected *counter-stories* that were then materialized as ultra-specific tools, such as Nikki's Tupperware grabber—made with a backscratcher to extend her reach, or Diana's bead collector (see Fig. 3.6)—a spatula wrapped in double-sided tape to collect fallen beads during crafting. Bennett and co-authors saw the design and making in this project to foreground difference and absence, to reveal voices, labor that had otherwise gone uncredited. Rather than universalizing unique standpoints, Bennett and co-authors call for foregrounding partiality and see the prototypes as capturing a moment: "*they are not meant to be comprehensive, authentic, or paradigmatic. Instead, they partner with their author for a brief and situated retelling.*" The ways of knowing that are brought together in this project are those of people with disabilities, and the practice of design. It reveals how the very qualities, knowledge and skills that are considered exceptional in professional design practice (creativity, inventiveness, and resourcefulness) are materially present in the lives of people with disabilities. And through this, what is known is not a "good design," but rather the potential of material things to function as situated retellings.

We have similarly created such situated retellings in our work on Weaving Stories (Oogjes and Wakkary 2022). In this project, we narrate actual events of design projects that involved weaving, including the warping of a loom, making samples, and testing and selecting materials. The retelling is done from different perspectives and with narrative strategies, to reveal unseen participation of nonhuman things such as tools, materials, and machines. For example, one story starts with narration from a Wensleydale sheep (a sheep known for its lush wool), to later include the speculative imagining of the thread's journey as it tangles, breaks and weaves—from the perspective of the thread. While the sheep was not physically in the room, and the threads did not actually have tiny human hands—bringing in these ways of knowing the world enabled us to attune to different aspects of the yarn. The qualities of wool are highlighted: its fibrous materiality makes it

Fig. 3.6 Diana's bead collector, a biographical prototype representing her story about spilling beads during crafting. Diana wrapped a spatula in double-sided Tape to extend Her reach from Her wheelchair and collect the beads from the floor. *Source* Courtesy of Burren Peil

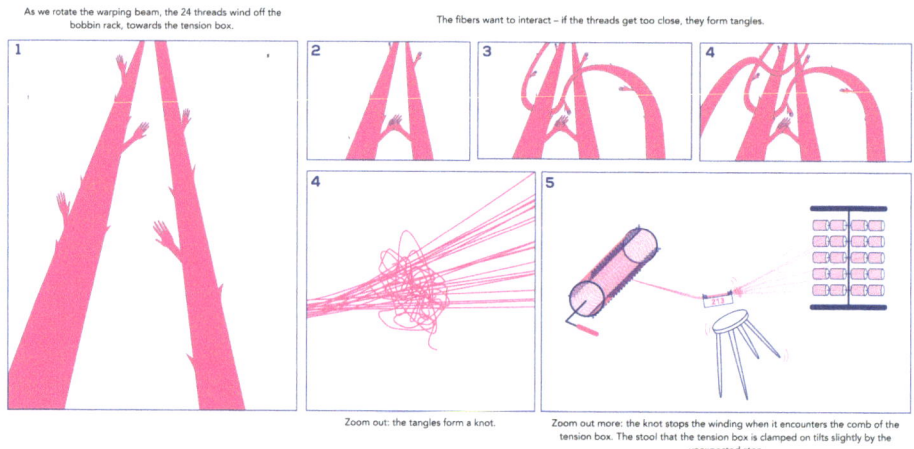

As we rotate the warping beam, the 24 threads wind off the bobbin rack, towards the tension box.

The fibers want to interact – if the threads get too close, they form tangles.

Zoom out: the tangles form a knot.

Zoom out more: the knot stops the winding when it encounters the comb of the tension box. The stool that the tension box is clamped on tilts slightly by the unexpected stop.

Fig. 3.7 A speculative re-imagining of the thread's journey during a warping process, based on true events. *Source* Doenja Oogjes

excellent to weave with, creating interactions between the warp and weft yarns to form a dense cloth. But this exact quality makes it harder to warp—those same fibers create opportunities for tangles and knots (see Fig. 3.7).

We note that nonhumans have always participated in design, yet often this becomes clear only when things do not go according to the plan of the human designer, say, when a thread breaks or a machine stops working. In Weaving Stories, we used speculative reasoning to bring forward different ways of understanding nonhumans of design towards other possible collaborations.

Examples like *RIOT, Biographical Prototypes* and *Weaving Stories* ask *"what if"* questions in material ways. What if menstrual health was prioritized in public washrooms? What if we prioritize stories, creativity and making practices of people with disabilities in design? What if we followed the journey of a thread, to engage with material agencies? There is an epistemological commitment in knowing through the material and the local, and through it alternative values and experiences—that were already present but obscured—are revealed. The focus becomes how the works speculate and make material, in the philosophical sense, other possible conditions that do or can exist concurrently in our present. This expands our attention from not only the technology itself but also the conditions that are in place for that technology to exist. In the examples in this chapter, there is an overarching interest in revealing the sociocultural, economical, systemic, infrastructural and moral conditions needed for these speculations to exist. Collectively these projects show how speculative frictions put forward different ways of knowing technologies, from maintenance infrastructures in the *RIOT* project, to everyday inventiveness in *Biographical Prototypes*, to nonhuman participation in *Weaving Stories*.

3.3 Taking Positions of Ethical Reflexivity

The speculative projects discussed so far in this chapter have shown that they prompt ethical reflexivity by displacing current assumptions and conditions, whether unquestioned technological developments in Alternative Presents, underexposed urgent topics such as air pollution in Southern California, or prioritization of concerns such as women's health and the creative agency and intellectual property of people with a disability. The speculations create a space to investigate the limits of the practices of design. This section turns to more examples of projects that explicitly leverage the power of the present to enable ethical (self)-reflexivity.

The *Hand's Up, Don't Shoot Glove* by Elizabeth Chin explores questions around the potential of technologies in Afro-diasporic spirituality and ways for technology to exist for preserving black lives. Embedded in the glove are a raspberry pi, accelerometer, and camera (see Fig. 3.8). When the hand wearing the glove is raised, it prompts a livestream of photos posted to the web. The project is created in response to police misconduct and violence and reveals the conditions in which an object like this is needed. It allows for speculation of on-body cameras and what is made possible through wearing them—and *who* wears them. The beadwork covering the glove is typical of Voudou and Santeria ritual objects, its imagery and colors referencing Santeria deity Chango, the king and dispenser of justice and Ogou, a Haitian and Caribbean figure representing iron—industrialization, technological infrastructures, and weapons. Through this, Chin reveals and reinforces connections between the spiritual realm and technology: "*Chango and Ogou operate* via *the glove as warriors for justice, fierce protectors who persevere against impossible odds*" (Chin 2022, 61). The material existence of the glove and design decisions of the beading colors and patterns reveal the obscured but existing relations between spirituality and technology. As such, it reveals a diverse epistemology. Yet, it also reveals the conditions of black life in America and prompts ethical reflexivity on relations with technologies such as cameras and surveillance and who it is directed to.

With the examples of speculating with the present, ethical reflections happen in the actual and material, and it is from this vantage point that researchers themselves are placed in a position to (self-)reflect. There is often a more personal research positionality that further allows for ethical reflection. As an example of this, *Watching myself watch birds* is an investigation into the personal birdwatching practices of a design researcher with the intention of unseating human designers from the center of focus. Through sound-recording, journaling, video collaging and reconstructing bird calls, Biggs reflects back on their increasingly intrusive, strange, and at times unpleasant encounters with birds and turns to theorist Julia Kristeva's concept of *abjection*. Abjection is a visceral response to a confrontation and rejection of borders of self, for example with nail clippings, pus or a dead body. For example, Biggs started dreaming about birds, and recalls being unable to escape their songs.

Fig. 3.8 Hands Up, Don't
Shoot Glove by Elizabeth
Chin. *Source* Courtesy of
Elizabeth Chin

In one of Biggs' video collages, a laughing track is overlaid on found footage of a
cardinal (see Fig. 3.9). Sounds of people chuckling and cracking up defamiliarize the
animal's innocence as it sits on a branch and creates an eerie, unsettling feeling. Biggs
describes this as an intentional shift: "*Somehow, for birds to be seen as beautiful and
natural preserves them from having an effect, to become closer to birds requires that they
somehow become strange, not neutral and affecting, not powerless. They need to somehow
become strange to become 'themselves' and become 'real' as I mentioned earlier*". The
work is not about birdwatching per se, but the birdwatching provides the conditions that
allowed Biggs to experience a process of forming a subject–oneself–that is increasingly
open and porous to broader ecologies. The speculative friction in *Watching myself watch
birds* lies in this ethical reflexive re-evaluation of self, questioning previous constructions
and boundaries of the human subject.

These works show that speculating in the present allows for encountering and negotiating ethical dimensions of design work. The positionality of the researcher is important,
and the use of first-person work is particular to speculating with the present.

3.4   Experiential Alternative of Things that Work Otherwise

Experiential alternatives are what makes speculation particular in design research. As
described in the introduction, each of the important qualities of speculation: leaps of
imagination, diverse epistemologies, and ethical reflexivity are methodologically enabled

Fig. 3.9 Birdwatching tools used by Heidi Biggs including field notes, birdwatching guide and Binoculars. *Source* Courtesy of Heidi Biggs

through the experiential alternative. In this section of the chapter, we turn to a discussion of what we see as commonalities in the making of experiential alternatives, through the examples of this chapter. We describe this as *things that work otherwise.*

Things that work otherwise are artifacts and systems that materially function in atypical circumstances following a rationale that is outside current norms. The experiential alternative counters the dictates of the role of technologies to reveal unseen or new conditions of present-day worlds.

As we have highlighted throughout the chapter, the material and everyday nature are important in countering these norms. The speculations are positioned as empirical, present-day artifacts that can and do function. The modalities range from utilizing everyday materials such as in Biographical Prototypes, low-fi deployable prototypes such as

RIOT, to higher fidelity, finished objects like the Tilting Bowl or the Hand's Up, Don't Shoot Glove. Perhaps most poignantly present in the Tilting Bowl project is the power of having these things work, otherwise. While it would be entirely possible to simply imagine a bowl that tilts, the speculation works because of its actual, physical presence—an approach that is called material speculation (Wakkary et al. 2015). The work going into making the Tilting Bowl as an experiential alternative includes the ceramic process, the electronic design and integration—common design processes enlisted to make a strange thing. The importance of making the Tilting Bowl work, including negotiations on the degree of its tilt, the sound accompanying it, the size of the bowl, making it food safe, are essential in allowing the speculation to reveal nuanced alternative realities in everyday life. This commitment to making absurdity work is also present in the crafted overlay of a bird video with a laugh track in Biggs' Birdwatching project, setting up a hotel system for plants and following through on details such as the plant care cards in Plant Hotel, or illustrating threads that have tiny human hands in Weaving Stories.

The experiential alternatives of the continuous present speculations reveal the present through the commitment of making the strange and absurd work. We will discuss in Chap. 5 the accompanying strategies and methods to these speculations including estrangement and empirical subjectivity. Next, we turn to the broader field of design research to see how the characteristics of speculating with the present are evident in research practices.

3.5 Continuous Presents in Design Research

This chapter sets out to position the continuous, unstable, and unresolved present as a site for speculation. These threads are clearly visible in the examples we have given—their material and empirical nature inevitably encountering friction. But like the previous chapter, this view is more prevalent in design research than commonly understood.

One example of this is Steve Jackson's concept of *broken world thinking* which assumes the world to be in constant state of breakdown, thus in need of upkeep—consider for example, the ubiquity of cell phone repair stands. Yet, Jackson argues, the worlds and labor of repair are rarely at the center of imagining or theorizing technology. The focus of broken-world thinking is in clear contrast with more dominant views of innovation, endless development and novelty (Jackson 2014). It is a leap of imagination within the present that centers on limitations, fragilities, and ongoing breakdowns of technology. The concept emphasizes work on maintenance and repair, mending and craft, bringing to the fore relationships of mutual care and responsibility between humans and things (see Fig. 3.10).

Imaginative leaps within the present can also be found in the classic and long-established practice of forming hypotheses. In essence, hypotheses are propositions or statements based on limited evidence—the incompleteness functions as a starting point

Fig. 3.10 Repairers at Gulistan underground market. *Source* Courtesy of Syed Ishtiaque Ahmed

for further investigation and testing. As such, leaps of imagination have always been part of hypothesis forming. Formulating a null hypothesis could be seen to rely on countering approaches in that it states the opposite of what is assumed to be true in order to reject and provide supporting evidence for the assumption. This can also be extended to lab approaches in research through design work in which design artifacts are seen as physical hypotheses (Stappers 2012) to test alternatives through experimental modes. For example, Joep Frens' work on rich interaction includes designing camera variations (rich, traditional and in-between) to statistically test an alternative interaction design paradigm (Frens 2006). As seen in Fig. 3.11, the study systematically analyzed four interface variations (varying use of buttons, sliders, etc.) and compared qualitative aspects (such as "beauty," "goodness," "ease-of-use") through user tests.

Diverse epistemologies of the present can be found in the rise of feminist perspectives in design research, for example, through topics such as women's health, as shown earlier in this chapter. Shaowen Bardzell further positions feminism (Bardzell 2010) through its critique and its central commitments including agency, identity, equity, empowerment, diversity, and social justice. Feminist epistemology (or standpoint theory) is an example of diverse epistemologies in that it takes alternative positions (traditionally, women's position) as starting points for research to reveal new avenues. Similarly, studies have drawn from ethnography utilizing standpoint theory and defamiliarization as a way to reveal

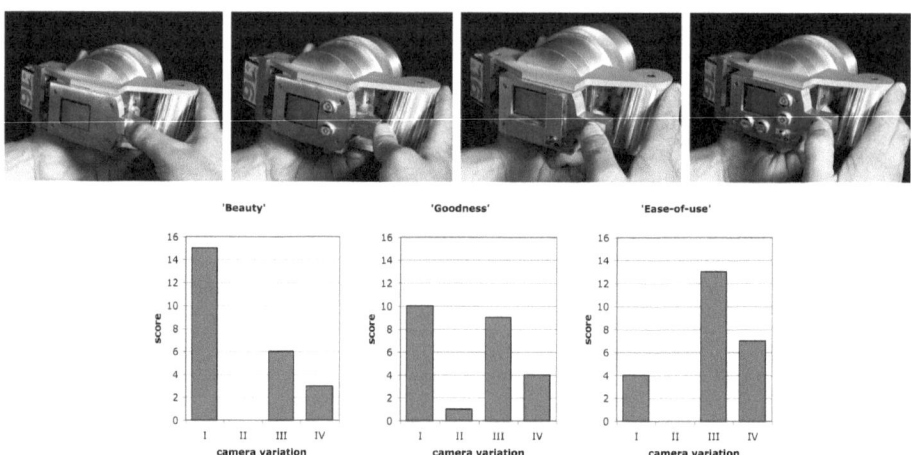

Fig. 3.11 Four variations of Joep Frens' rich interaction cameras, tested on beauty, goodness, and ease-of-use. *Source* Courtesy of Joep Frens

alternatives that are already present in the everyday (Bell et al. 2005). Bell and Dourish reflect on the shed as an instance to investigate the edges of domestic life (Bell and Dourish 2007). Through this, they aim to broaden ubicomp's understanding of the home, specifically, the ways in which it is socially and culturally constructed and enacted. The authors position the shed as a space that resides on the boundaries of what is considered domestic, a "liminal space" that can be taken to understand domesticity differently. The shed is considered a masculine space (a place for DIY, dirt and messiness), a dangerous place (where things are such as rat poison and power tools are stored that do not have fit in the home) and a transitional place (where things move in and out of the domestic sphere, for example through repair). Through this analysis, Bell and Dourish establish a broader conceptualization of "the home" that reveals questions of safety, cleanliness and gender—opening up new lines of research on the home including studies on different forms of domesticity—such as co-housing (Jenkins 2017; Shin et al. 2019)—and activities that constitute homelife that had previously remained under-examined–such as "pottering," unplanned activities supporting the upkeep of domestic life like tidying, gardening and organizing (Taylor et al. 2008).

Earlier, we described how Wu and Koskinen (2022) utilized breaching experiments with the Plant Hotels. Ethnomethodology and its methods have long been part of the tradition of design research. As we explained earlier, breaching experiments provoke—as in make visible—the taken for granted and invisible organization of everyday life (Garfinkel 1967). Andy Crabtree positions breaching experiments as a way to study in the "absence of practice"—arguing that technological development often precedes or overlooks grounding in social organization and everyday routines as they do not exist yet (Crabtree 2004).

Breaching experiments have been widely adopted in design research and often make use of familiar settings to employ novel designs and technologies "in the wild," drawing from the familiar to encounter friction. For example, by making visible the system's shortcomings, unintentional consequences, and surprising responses. A well-known study utilizing breaching experiments is the Ghost Driver field study, in which researchers created a car seat costume for the driver to wear to observe pedestrians' reactions to self-driving cars (see Fig. 3.12). While many of such studies focus on future possibilities and technological innovation, the roots of the method emphasize a focus on revealing what is already present in the everyday, and enable ethical reflexivity through material encounters: "*I have found that they produce reflections through which the strangeness of an obstinately familiar world can be detected*" (Garfinkel 1964).

Fig. 3.12 The Ghost driver study setup, with a seat cover costume worn in the car during the experiment, to observe people's reactions to a self-driving car while crossing the street. *Source* Courtesy of Dirk Rotherbücher and Colleagues

3.6   Summary

To summarize, in this chapter, we focused on speculation that engages with the present. We characterized this as a revealing of everyday conditions that (could) generate alternatives. We drew from approaches such as Alternative Presents, material speculation, and the experiential alternative of *things that work otherwise*. Like the previous chapter, we applied our framework of leaps of imaginations, diverse epistemologies, ethical reflexivity and experiential alternatives to draw out how these speculations work. Lastly, in our continued argument of positioning speculation's importance to broader research in design, we brought in work that has been ongoing in the field that shares aspects of our framework as they materialized in our examples.

Continuous Pasts

In the Salish Sea, just off the coast of Vancouver in British Columbia, where we are currently writing, is Penelakut Island, named after the Puneluxutth' peoples or First Nation. The origin of the Puneluxutth' people is told through Indigenous storytelling or a creation story. Creation stories are enduring, passed on orally through generations (Kovach 2010). In one telling of the story, the Puneluxutth' elder Mary Rice tells the white settler journalist Cryer (2008) that the meaning of "Puneluxutth'" in Hul'qumi'num, the language of Puneluxuth', is "two logs covered in sand." She recounts that on the island where there were "only the birds and animals," there was near a spring "two great cedar logs, lying on the ground and half-covered with sand." The sun shone on one of the logs and "Bye and bye the bark on this log began to crack…Suddenly there was a loud crack! The bark split open, and out came a little man!" Soon after, "he saw the sand between the two logs open…and out came a little woman!" (Cryer 2008, 126). The two stripped bark of nearby cedar trees and built themselves a house that they named after the two logs. "So those two first people called their house 'Puneluxutth'" (Cryer 2008, 127).

The island was renamed and reclaimed in 2010 from its colonial heritage in which it was formerly known as Kuper Island. In 1851, a British Royal Navy frigate named *HMS Thetis* occupied and claimed the island, naming it Kuper after the captain of the frigate, Augustus Leopold Kuper. A nearby island was renamed Thetis, after the frigate. The colonial occupation asserted its own origin story of "discovery," eliding the truth of occupation and theft of lands, peoples, and resources. It is the origin of past dark stories that include the dispossession of the lands of indigenous peoples throughout the Salish Sea that fomented a devastating massacre of the Puneluxutth' and British Navy attacks of the island in 1860s; the founding of a residential school in 1890 by the Roman Catholic

Church and the Canadian Federal government to forcibly deracinate and assimilate indigenous children until its closing in 1975, a period in which over 120 indigenous children were confirmed to have died; and stories through to today that make clear the ongoing need to decolonize.

The dominant origin story of the past shapes the present, often forcibly. Despite this, the past is not fixed, rather it is open to negotiation, interpretation, and the interweaving of multiple origin stories. As such, the past positions us in the present. Speaking as the first author, I am a second-generation uninvited settler to Canada. I am not White-European but the son of a Sulawesian father and Javanese-Sundanese mother who settled together in Canada in the 1960s. And I married into a Dutch family from North Brabant in the Netherlands who also settled in Canada at the same time. One part of the maternal side of the family, originally German, migrated to the Netherlands while the other to Britain. In the line of descendants on the British side is Augustus Leopold Kuper, the captain of the *Thetis,* making him my ancestral in-law. The past positions one in the present, entangles us in ways that are lively, including in dark ways, necessitating investigations, formulations, and reformulations. The past opens one to multiple perspectives not always of our choosing. As an Indonesian with a Canadian passport, at border crossings, I can sometimes be seen as strange if not suspicious. As the feminist philosopher Sara Ahmed describes, as a British citizen with a Pakistani father, a border crossing can open one's past into a provisional and speculative state, requiring further questioning of where one is really from or a past to be interrogated such as "where is your father from?" (Ahmed 2006). Ahmed describes this as being made a stranger to oneself. Estranged by a past that becomes negotiable for some to make one "out of place" with their present position (Ahmed 2006, 141).

If the past is open to different perspectives not always of our choosing, it is also open to our intervention. In *Lose Your Mother*, Saidiya Hartman retraces stories of the Atlantic slave trade on a year-long journey along a slave route in Ghana. In a careful consideration of whose perspectives to share, and which stories to bring forward, Hartman's work exists somewhere between analytical, autobiographical, historical, and fictional. She actively returns to, recognizes, and uplifts untold histories and brings them into the present—an active commitment of diversity by portraying agency of Black people in history, however, without denying the terrors or falling into hopes of resolution. Her work is that of reconsidering the archives of slavery, to go beyond records of property and death, and through stories to represent the lives and humanity of the unknown, nameless, and forgotten. It is also an ongoing quest into how one returns to the past and writes about horrors, violence, and death.

In one of the stories in *Lose Your Mother* (Hartman 2007), Hartman writes of the murder of a girl on a British slave ship that was followed by trials and debates over the abolition of the slave trade. In her original account, Hartman briefly mentions another captive girl on the ship, Venus, who appeared in the trial records. The essay *Venus in Two Acts* (Hartman 2008), is a self-critical return to this girl—reconsidering which voices

Hartman left out in her own work. Hartman first uses the term *critical fabulation* in this essay, which she categorizes as a writing practice that reworks the building blocks and basic elements of a story.

> By playing with and rearranging the basic elements of the story, by re-presenting the sequence of events in divergent stories and from contested points of view, I have attempted to jeopardize the status of the event, to displace the received or authorized account, and to imagine what might have happened or might have been said or might have been done. (Hartman 2008, 12)

The practice of critical fabulations has precedents in Hartman's own work as well as that of, for example, Toni Morisson's *Beloved* (1987), which finds its origin in a newspaper clipping. For Hartman, *Venus in Two Acts* emerged from a self-reflection, one in which she criticizes herself for her silence around Venus in *Lose Your Mother*. In *Venus in Two Acts*, she asks: what else is there to know? What other fates are there? And takes imaginative leaps in filling in blanks of Venus' story with speculative details such as a friendship between the two girls.

In this chapter, we shift attention toward critically re-interpreting the past but not to give a historical account, to give a speculative account.[1] This speculating on the past makes the past provisional in a critical sense. It opens it to reveal hidden realities that shape current-day realities or offer new possibilities for the present or future. Such speculations are offered here as modes of inquiry in design research. In what follows, we focus on speculative practices in feminism, Afrofuturism, more-than-human epistemologies, disability studies, and decolonization as they relate to design and technologies. Like the previous chapters, we examine the workings of speculative reasoning through the framework of *leaps of imagination, diverse epistemologies, ethical reflexivity*, and *experiential alternatives*.

4.1 Leaps of Imagination into the Past to Trouble the Present

Leaps of imagination that look to the past can reconfigure the present. The Bikes and Bloomers project led by Jungnickel (2018) reconstructed a collection of Victorian era cycling skirts (see Fig. 4.1). She and her team researched archival materials and patents to find cycling skirts they could design and fabricate today. For them, the skirts were political acts that countered the constrained identities and agency of women. For example, the skirts could be worn daily to meet gender conventions while also allowing conversion into a cape or other types of garment to ride a bike safely. These inventions were subversive, they supported women to actively circumvent societal norms and impositions while still

---

[1] This resonates with what's known as "subaltern" or "insurgent" histories in subaltern studies, an offshoot of postcolonial studies, focused on the subjugation of South Asian culture and societies by colonialism or other forms of oppression.

Fig. 4.1 Kat Jungnickel (center) and some of her Goldsmiths Research team demonstrating patented Victorian cyclewear inventions they reconstructed for the Bikes & Bloomers Research Project. www.BikesandBloomers.com. *Source* Charlotte Barnes, www.CharlotteBarnes

participating within society. For Jungnickel, the patents show how the question of successful design and technology is reconfigured in these skirts. Their success is to be hidden rather than "told through loud or triumphal narratives" or as further stated "These inventors put in an awful lot of work to *not be seen*" (Jungnickel 2018, 253). For Jungnickel, the project "is a classic feminist reclamation project" (Jungnickel 2018, 6).

As a speculation of the past, Bikers and Bloomers purposefully aims to reconfigure the present. The remaking, wearing, and cycling with the historical clothing in our present-day reclaims the successes of these inventions. The inventors and modes of invention are also resituated to remake present-day narratives of technological success. Making space for subversive designs that support the politically marginalized of the past and expose current views of technology and invention today. The work asks not only what of the past do we not know or has been left out but can we read design and technological inventions as a "acts of political resistance, contestation, or subversion and what kinds of citizens are made possible or reimagined…" (Jungnickel 2021, 15).

Imaginative reenactments of the past include radical reinventions of present-day technologies and artifacts that in themselves serve as political interventions. In 2021, the Museum of Metropolitan Art in New York City, led by the guest curator Hannah Beachler, made steps toward reparation and self-intervention (Alteveer et al. 2022). The museum installed *Before Yesterday We Could Fly: An Afrofuturist Period Room* that speculatively and symbolically creates the home of a Black Seneca Village resident as it might exist today (see Fig. 4.2). Black Seneca Village was a free Black community in New York City that in 1857 was expropriated and destroyed to make way for Central Park. This destruction was fueled by a campaign founded on racist stereotypes that the village was poor, unkept, and unhealthy (Tillet 2021). Installed in the *American Wing Period Rooms* of the museum, like other period rooms, the installation is an immersive domestic space populated with everyday furnishings and objects to create the illusion of a historical day in the life. Unlike other period rooms in the museum, this Afrofuturist period room rejects a single history (told as the dominant colonial and White history of America) to embrace the interweaving of the past, present, and future, emblematic of Afrofuturism, as well as African and African diasporic beliefs. As Alteveer and colleagues write, the installation is an "informed speculation about the past with an ingenious design for the future…thus provides both a visionary leap forward and a radical counterpoint to so much historical oblivion" (Alteveer et al. 2022, 12).

Fig. 4.2 The before yesterday We Could Fly: an Afrofuturist period room at the Metropolitan Museum of Art. *Source* Art Resource

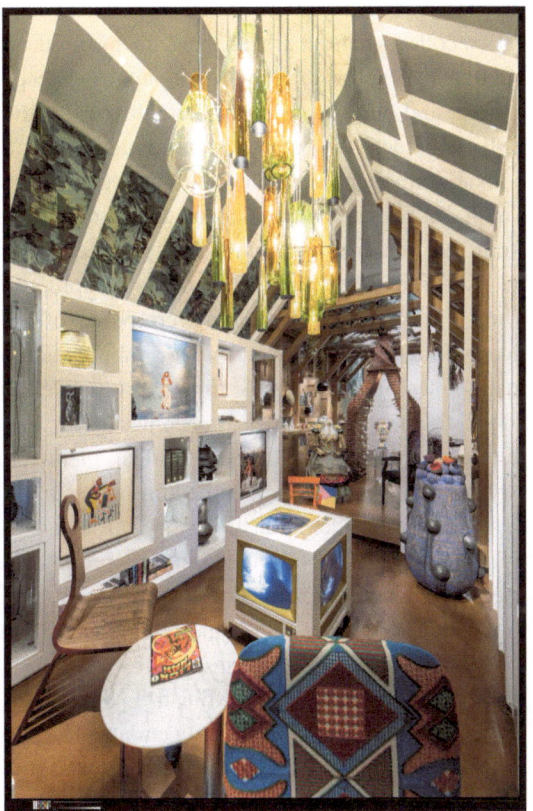

The name of the period room is inspired by the children's book author Virginia Hamilton's retellings of the legend of Flying Africans (Hamilton 1993). An oral tale of Africans who escaped their enslavement in the New World through the power of flight, able to take to the air like free birds and fly home to West Africa. These layers of speculation take the form of the transdisciplinary powers of Black imagination, creativity, and self-determination expressed through Afrofuturism. The room is furnished with rich entanglements of Bamileke beadwork, nineteenth-century ceramics to contemporary art and design (Alteveer et al. 2022). The curators commissioned work especially for the room, including a collage by Nijdeka Akunyili Crosby that interweaves imagery of Seneca Village and Black New Yorkers of the time with floral patterns based on the okra plant imported with slaves to America; and Jenn Nkiru's five-sided television showing a film of archival footage along with re-enactments of residents of Seneca Village dining together. This leap of imagination revives an erased past to make it full of possibilities again. It further intervenes through a new language to question and challenge the dominant telling of the past to make them equally provisional and lessen their hold on the present.

Speculative reasoning that revives the past can also include ecological histories and recognition of multispecies histories. Rachel Clarke's Ministry of Multispecies Communication collects data in urban environments on the well-being of species other than humans (Clarke 2020). The agency is fictional; it is a leap of imagination that is embodied, participatory, and performative. As Clarke explains, the Ministry conducts experimental urban walks in which participants, recruited residents of the neighborhood, wear masks of nonhuman animal species that in the past claimed that neighborhood as their habitats (see Fig. 4.3). The masked walks aim to be transformative by facilitating play and reflection from the imagined and embodied perspective of a bird or a fox, depending on one's mask. The masks are made from photographs of species from the location of the walks. The species chosen by Clarke were either of the past or present that mostly go unnoticed. Booklets are provided with prompts and space for participants to draw maps and make notes along the walk from the imagined nonhuman perspectives. This forms the "data" for the "Ministry" and becomes a way to explore alternate technological configurations of the city. For Clarke, this leap of imagination recasts conceptions of the smart city, the present and future embedding of technologies into urban spaces. The speculative reasoning reconsiders that trajectory through past or diminishing species and ecologies by uncovering actors of the past and present that are overlooked—namely nonhuman species—in the design of cities. At its most speculative, some participants were given masks that they were told could communicate with disappeared or disappearing species to bring them back to the city.

Within the fiction of the Ministry, participants were treated as "trainees" who would become part of the agency to collect data. The speculation enabled participants to engage in the rethinking of the multispecies histories and hence the future of the city along the more than human aims Clarke imagined. This includes seeing cities and technologies in ways that decenter human agencies to include nonhuman agencies; exploring through the

Fig. 4.3 Participants of the Ministry of Multispecies Communication on an urban walk in London wearing masks of Nonhuman animals. *Source* Courtesy of Rachel Clarke

lens of more-than-human temporalities that can be much shorter or longer scales than human temporalities; and incorporating other forms of intelligence that are more-than-human (Clarke et al. 2019).

In summary, these leaps into the past utilize imagination to make history provisional, and in a sense expand the past and reveal its multiple histories such as the overlooked feminist politics of the Victorian era, the erasure of Black lives and ways of living, and the cohabitation of nonhuman species. Speculations that spawn new trajectories that lead us back to the present to tackle salient and intractable issues.

4.2 Diverse Epistemologies of Repair

Speculations of the past open to different ways of knowing and acting on that past. The Karamū Stream, a river in New Zealand, was a site for a co-design project on water governance that for Manuhuia Barcham also became a research inquiry into a decolonial imaginary (Barcham 2022). The idea of a decolonial imaginary is drawn from Pérez (1999). It is an imaginary in opposition to what she identifies as colonial imaginaries that enforce hegemonic powers and rely on narratives of universality, modernity, and progress. Barcham adds that design plays an important role in decolonial imaginaries through informal and minor structuring to create possibilities, in contrast to grand social engineering of colonial imaginaries. The Karamū Stream is traditionally and currently a significant food-gathering site for the Ngāti Hori, who is a local hapū (clan) in Hawkes Bay, New Zealand. According to Barcham, the river is known by Ngāti Hori as Ngaruroro-Waimate or the Ngauroro of the sick water. This naming speaks to the ongoing devastation of their ancestral waterway since colonialism that has dramatically reduced water levels and diminished biodiversity. Given the deep relational interdependence between the Ngāti

Hori and the river, the traditional ways of living of the Ngāti Hori have been equally devastated.

Barcham, a bi-cultural *Māori* from New Zealand, was engaged by the Ngāti Hori in a co-design process to have a voice in the Karamū Stream Enhancement Project that was part of a resource management plan by the Hawkes Bay Regional Council. The outcome of the co-design was a cultural map that became a material and provisional space for multiple ways of knowing, including "Western science (e.g., the hydraulic geometry of channels at various points in the surrounding terrain) and others from mātauranga Māori (Māori knowledge) (e.g., the stories of the local taniwha (ancestral guardian spirits) associated with the river)" (Barcham 2022, 11–12). In addition, the co-designing included walks along the river that elicited and situated stories of the river as told by the different stakeholders. These stories became part of the mapping process. A process in which diverse epistemologies were "woven together—more tightly where different aspects overlapped in meaning and more loosely where they differed" (Barcham 2022, 12). This brought together different ways of knowing to find congruence in key areas and marked differences in other areas. Barcham makes clear though that the group prioritized Ngāti Hori knowledge and ways of being, ensuring their voice and in keeping with a decolonizing approach. As a result, the cultural map deeply informs decision-making by making the practice of *rahui* (traditional restrictions) equal to, if not greater, concern than the environmental science of seasonal reproduction patterns of fish species and flow rates of the river. The speculative reasoning by Barcham reclaims indigenous pasts as lively, extant, and critical to enact change now.

We have hinted throughout that the underlying temporal structure of the book would break down into a fluid and non-linear understanding of time. Further, temporal phases of past, present, and future hold multiple material realities or ontologies that intertwine but also, as we have seen, are overlooked or erased. Making the past provisional is to make it lively again as with decolonial imaginaries, making it present and at hand. The essay *Making Kin with Machines* (Lewis et al. 2018) brings together authors who draw on Hawaiian, Cree, and Lakota cultural knowledges to conceive of frameworks by which Artificial Intelligence (AI) could be seen as kin, belonging to one's family or other circles of relationships. The authors speak as members of these different ways of being rather than representatives or spokespeople of indigenous epistemologies. Nevertheless, as a set of individual explorations, each brings forward different material and cultural relationships from which to situate AI, sharing the broad questions of "how do we as Indigenous people reconcile the fully embodied experience of being on the land with the generally disembodied experience of virtual spaces? How do we come to understand this new territory, knit it into our existing understanding of our lives lived in real space, and claim it as our own?" (Lewis et al. 2018, 3–4).

In keeping with diverse ways of knowing and being, the essay is expansive. For example, Noelani Arista writes that the cultural concept of *aloha* for the *Kānaka maoli* (Hawaiian people) "is a robust ethos for all our relationships, including those with the machines we create" (Lewis et al. 2018, 6). Archer Pechawis, as nēhiyaw (a Plains Cree person), accepts that many Cree people would reject AI from their kinship circle, yet pursues the idea that "even with conditional acceptance of AI as relations opens several avenues of inquiry. If we accept these beings as kin, perhaps even in some cases as equals, then the next logical step is to include AI in our cultural processes" (Lewis et al. 2018, 7). Lastly, in the last of the reflections within the essay, Suzanne Kite invokes wakȟáŋ from Lakota knowledge, referring to that which cannot be understood, viewing the ontological status of AI as an irreversibly interconnected whole:

> I am not making an argument about which entities qualify as relations, or display enough intelligence to deserve relationships. By turning to Lakota ontology, these questions become irrelevant. Instead, Indigenous ontologies ask us to take the world as the interconnected whole that it is, where the ontological status of non-humans is not inferior to that of humans. Our ontologies must gain their ethics from relationships and communications within cosmologies. Using Indigenous ontologies and cosmologies to create ethical relationships with non-human entities means knowing that non-humans have spirits that do not come from us or our imaginings but from elsewhere, from a place we cannot understand, a Great Mystery, wakȟáŋ: that which cannot be understood. (Lewis et al. 2018, 11)

The speculative engagement of traditional practices and ways of knowing with futures is an act against defuturing and exclusive ownership of the future by select histories.

Seeing the past as provisional opens the present to being shaped by the hidden narratives and counter stories of the past whether it concerns environmental management or AI. Rua M. Williams and colleagues build on the role of counternarratives to counter stigmatic and oppressive cultural assumptions often found in disability research (Williams et al. 2023). Specifically, they seek a critical yet productive approach to research studies that intervene to create normative behavior change amongst disabled people, implicitly privileging able-bodied biases. The authors propose *counterventions* to critique such studies utilizing counterarguments that are community informed and seek to produce imaginative alternatives. Informed by queer theorist Eve Kosofsky Sedgwick's *reparative reading* method, the goal is to prioritize understanding by integrating disparate elements to create positive action, moving beyond a simple critique of latent or blatant dangers and harms (Sedgwick 1997). Like reparative reading, William and colleagues accept a past that "could have happened differently from the way it actually did" as the way to act toward a more liberatory future (Sedgwick 1997, 25).

Williams and colleagues offer counterventions of studies of autism that present a "single story" of the need to intervene to produce normative neurotypical outcomes. They focus on the tensions or conflicts researchers report of participant responses to formulate counterventions that lead to new opportunities within the research. For example, they seek to repair a study focused on visualizing the prosody of a neurodivergent participant's voice to encourage them to use less of a flat tone. The reparative move is to visualize the voices of those around the participant as an alternative. Similarly, another study targeted neurotypical norms of personal space or proximity aiming to encourage neurodivergent participants to keep a certain distance from others while interacting. The reparative alternative is to utilize the system to make neurotypical participants aware of alternative interactions without judging intentions of those involved. In large part, the counterventions bring to light obscured neurodivergent norms as a matter of diversity and shift the onus and burden for change from disabled peoples to collectively negotiating modes of interaction among all stakeholders.

In summary, speculative reasoning makes concrete different ways of knowing the past or extant knowledge that is kept from view through giving presence to counternarratives or alternative stories. This approach abandons assumptions of a universal history or a single story to create possibilities and opportunities to reshape current understandings and actions.

4.3   Making Stories of Ethical Reflexivity

In the works discussed in this chapter, researchers, artists, and designers purposely intervene with counter-histories to make the past provisional, open to inquiry and redirection. This reflexivity materializes erased and obscured histories in part to reclaim them but more so to critically reflect on today. The Bikes and Bloomers project (Jungnickel 2018) discussed earlier reclaims a past mode of critical creativity that asks us to question today's "triumphal narratives" of technological invention and discovery to make space for inventions that are subversive and offer political resistance. The project *Making Core Memory* is an example of Daniela Rosner's approach of feminist correctives,[2] setting out to investigate, redefine and re-root design in ways that are more inclusive of methods, people, contexts and histories that have been marginalized or obscured (Rosner 2018). The project examines lineages and gender dynamics of craftwork and its contribution to computing, centering stories of the women who wove the cores for the Apollo 8 spacecraft in the 1960s (Rosner et al. 2018). The making of core memory consisted of threading metal wires through the cores, performed by women including many women of color and with roots in Navajo weaving. The practice was often dismissed by Apollo engineers as feminine craft, referring to the weavers as Little Old Ladies, or L.O.L. in short. In the Making Core Memory project, Rosner and colleagues collaboratively weave new core memory in workshops with educators, technologists, students, and historians. The woven patches are

---

[2] Daniela Rosner uses the term *critical fabulations* for her approach of feminist correctives in design. We find it important to make clear that this should not be confused with Sadiya Hartman's use of the term. As we described earlier, we see Sadiya Hartman's critical fabulations as an investigation of the forgotten, obscured, and marginalized voices, also those in her own writing (for example, in the Return to Venus, as described in the introduction to this chapter). See the post "Critical Fabulations Beyond Design: On Being Informed by Saidiya Hartman" (https://medium.com/@danielarosner/critical-fabulations-beyond-design-on-being-informed-by-saidiya-hartman-3509bd6e4f43) for Rosner's reflections on her use of the term in connection to Sadiya Hartman's work.

Fig. 4.4 Core memory and the associated twitter account that posts quotes and updates of the patch-making process. *Source* Courtesy of Morgan Ames

connected on a quilt with a server that tweets quotes of the Little Old Ladies, for example, details on the making of the cores, expressions on the working environment, and instances of interactions with NASA engineers (see Fig. 4.4). The Making Core Memory project reveals the perspectives of the Little Old Ladies by retelling (part of) their stories. Its ethical reflexivity lies in the repositioning of craft and women's work by weaving the cores, enabling workshop participants to experience the required detail and precision of the work that was underappreciated in its own time.

Laura Forlano and Megan Halpern speculatively engage the past to reflect on labor through *counterfactual actions* (Forlano and Halpern 2023). Like our discussion, history is seen as a speculative space to reveal political, ethical, and social justice alternatives that can inform future action. The counterfactual is common in philosophy and literature, as a what-if speculation of the past to inquire on alternative narratives of the present or future. Counterfactual actions are a methodology for participatory and community-based design. They employ modes of speculative reasoning from thought experiments to roleplay to gameplay that "engage deeply" with history but are not necessarily bound by historical facts. The aim is to reveal or create new socio-technical interdependencies that shape identities and processes. For example, in a workshop on the future of work the authors invited over 20 labor and technology activists, designers, historians, and other scholars to critically reflect on labor, technologies, and socio-economic conditions. To move past present-day assumptions and demands toward longer-term thinking, the authors created a game that required participants to invent counterfactual actions related to labor and technology, such as worker campaigns, across 3,000 years of history. For example, in playing the game a team of participants might be challenged to "plan a collective action with fellow workers" when on a space on the board marked "pyramid builder" in Ancient Egypt (see Fig. 4.5). Making the past provisional and speculative productively defamiliarized

Fig. 4.5 The "Reimagining Work" project, a speculative board game about technology and labor. This timeline was divided into five Eras. Each Era had six spaces with an identity. Teams Would Roll a Die, Land on a space, and draw a card. *Source* Courtesy of Laura Forlano

assumptions of labor, technology, and futures. The game provided historical insights into labor activism that inspired participants; discussions emerged about the persistent role of time-based technologies in regulating labor through to today; and longer and diverse perspectives emerged that spurred on creative use of technologies for labor dissent and collective organization.

In summary, these moments of ethical reflexivity arise from directly speculating on the past, using the past to redirect efforts or see anew in ways that attend to purposely obscured or unseen contributions and labor actions.

4.4   Experiential Alternative of Materialized Stories

As a reminder, speculations require experiential alternatives to methodologically enable leaps of imagination, epistemological diversity, and ethical reflexivity. We have concluded each chapter by explaining how speculative reasoning is enabled through the making of experiential alternatives. And so, we now turn to a discussion of what we see as a commonality within the experiential alternatives in the examples of speculation in this chapter. We describe this as *materialized stories*, experiential alternatives that seek to remake and reenact the past critically and creatively.

Materialized stories are reenactments or the making of artifacts of the past to open history to critical inquiry by making it lively, vivid, and provisional. The experiential alternative counters the past as settled and singular to reveal obscured, erased, or ignored histories to change the trajectories of the present and future. It is clear that the materializations have a wide range of expressivity and use of materials and technologies when comparing the detailed reenactments of the clothing in the Bikes and Bloomers project (Jungnickel 2018) to the imaginative re-interpretations in the patch and quilt making of Little Old Ladies (Rosner et al. 2018) (see Figs. 4.1 and 4.4).

The importance of expressivity is evident in *Before Yesterday we Could Fly*, in which there is a multiplicity of creative approaches to seize the opportunity to shape the speculation (Alteveer et al. 2022) (Fig. 4.2). The materialized presence of the "period room" powerfully diffracts modes of inquiry into matters of race, history, and reclamation. The speculation invites the museum visitor to an imagined performance of a "day in the life" of a Seneca Village resident. Given the creative and critical reenactment of the provisional past, the visitor cannot engage the performance without the weight of historical erasure and racism.

The liveliness of these provisional pasts is as performative as they are material. The materializing of past animals in the form of a mask is made more lively in the wearing of the mask to embody a nonhuman perspective (Clarke et al. 2019) (Fig. 4.3). The performative qualities of the experiential alternatives create a "co-presencing" of histories and different positions that form the crucible for countering the present or dominant narratives.

Similarly, Bikes and Bloomers (Jungnickel 2018) underscore the co-presence of past and present by wearing and riding with the clothing to assert the relevance today of this revealed history (Fig. 4.1).

The experiential alternatives of continuous past speculations critically open the past through expressive and performative materializations. In Chap. 5, we will discuss the accompanying strategies and methods for these speculations, including countering and counterstories. Next, we turn to the broader field of design research to see how the characteristics of speculating on the past are evident in research practices.

4.5   Continuous Pasts in Design Research

Despite the future facing imperatives of much design research, the importance of the past and opportunities within the past is well recognized. In line with speculation as we see it, design research has adopted in different measures the recognition that dominant histories obscure (or erase) other histories; and dominant imaginaries diminish or ignore longstanding practices and prior knowledge. In Chap. 2, we discussed critical literature reviews as creating leaps of imagination into future research possibilities. Here, we can see how a complementary move in literature reviews is to see past research, a field's history, as provisional. As researchers we structure, restructure, and argue for narratives of past research to reveal obscured or overlooked possibilities that can be addressed in future research. Ethnography and ethnomethodology have served as a counter to untested beliefs in technological innovation and its imposition of a dominant understanding of human behavior. This is a long tradition that includes Lucy Suchman's pioneering observations of scientists and engineers using a new copying machine at Xerox Parc in the early 1980s (Suchman 1987). Her ethnomethodological study reveals the divergence of intelligibility between longstanding practices of people and the formalized reasoning and constraints of technological functionality embedded in interactive machines like a copier. Unseen by the engineers and designers at the time was a deep asymmetry in the interactions between people and new machines.

Another long-standing practice of design, participatory design, can be said to take this divergence between existing and past practices of people and imposed new practices of technologies as a core challenge. Pelle Ehn in his doctoral thesis identified the role of design to work through the "dialectics of tradition and transcendence" (Ehn 1988, 7) and to do so through the inclusion of stakeholders in the designing of a new system. Participatory design developed many key methods to design democratically or to co-design. An early example of such a method is what Ehn called *design games*, a way of making potential future designs both provisional and accessible to the non-designers whose lives were to be impacted by a new system. For example, Ehn describes co-designing with Graphic Union Workers in Sweden in the early 1980s a computerized text and imaging system that would profoundly transform work practices from largely analog to digital, requiring new tools, workspaces, as well as new competences and skills. In a design workshop, participatory designers and workers in a design game created cardboard mock-ups of workstations and printers to move around the offices and assist in enacting new routines and material practices (Ehn and Kyng 1992). The design game is speculating that bridges past practices and new understandings in ways that materially bring together tradition and transformation into a shared dialogue (see Fig. 4.6). Like many of the examples in this chapter, participatory design brings together multiple stories and practices into a productive dialogue as an emancipatory practice that in PD is grounded in democratic inclusion of workers and labor into the design of technological systems.

Design researchers have also engaged explicitly with history as it relates to technologies. For example, Ylva Ferneaus and co-authors draw threads from a reconsideration of the Jacquard loom to bring new insights to the field of tangible and physical interaction design (Fernaeus et al. 2012). The weaving loom is often considered one of the predecessors of computer science (through its binary configuration in the use of punch cards). Here, it is unpacked further how the complex machine has a quality of graspability—a concept that is prevalent in the design of tangible artifacts. The study demonstrates how a careful unpacking of a historical design can bring nuanced perspectives and complicate

Fig. 4.6 A cardboard mock-up of a laser printer that in a design game allowed stakeholders and designers to "Make-Believe" a system that they could enact practices with but also easily change and redesign. *Source* Courtesy of Morten Kyng

distinctions and challenge assumptions on current themes within design (such as embodied interaction, sustainable interaction design, and material-focused interaction design). In a more recent example, Samar Sabie and colleagues reexamine Luddism as an "unmaking" movement that is seen as a current approach to undermining and countering oppressive structures (Sabie et al. 2023). The Luddism movement was a collective resistance to the industrialization of textile industries in England in the 19th century that is characterized by the destruction of textile machines as political protest. Sabie et al. revisit Luddism to uncover alignments with current work on unmaking and social justice.

These examples are part of ongoing engagements with the past in design through historical research to explore "paths not taken" and foreground untold histories (Soden et al. 2019; Balka and Wagner 2021; Mensonge 2023). Specific foci include archival studies (Turner et al. 2021; Lindley et al. 2013; Seberger 2021), and investigations of steampunk maker practices as they relate to design fiction (Tanenbaum et al. 2012). For researchers, this opens the door to productively reimagine the technological artifacts produced within a longer temporal frame. For example, there is a large body of work exploring concepts such as heritage (Loonker et al. 2022; Schofield 2014) and heirlooms (Banks et al. 2012; Baytaş et al. 2018; Jung et al. 2011; Odom et al. 2012) historic data as a design material (Fedosov et al. 2018; Holstein et al. 2020; Pschetz 2008; Thiry et al. 2013; White et al. 2023) and the passing on of personal (digital) belongings, family history and remembering the departed (Elsden and Kirk 2014; Gulotta et al. 2013; 2015; Kirk and Sellen 2010; Massimi et al. 2011).

In these practices, speculative reasoning may not be foregrounded, but the characteristics are clear. In the language of speculation, some design research practices similarly see it productive to make the past contingent and provisional to inform the present and future.

4.6   Summary

In this chapter, we focused on speculative reasoning that makes the past a provisional site for new inquiry and productive possibilities. We characterized this aspect of speculation as countering dominant or single stories to reveal marginalized, obscured, or even erased narratives that are critically relevant to today. This chapter traced speculative practices through feminism, Afrofuturism, more-than-human epistemologies, disability studies, and decolonization. Again, as with the previous chapters, we utilized our framework of leaps of imaginations, diverse epistemologies, ethical reflexivity, and experiential alternatives to show the forms of speculation and how the past can be generativity and critically engaged with. We also argued that similar considerations of the past and speculative reasoning have been important to design research and design more broadly and for some time now.

The Practice of Speculative Reasoning

5

Our motivation in writing this book is to claim that speculation is central to design research through what we call speculative reasoning. Speculative reasoning is the creative use of propositional knowledge that applies different ways of knowing to counter dominant modes to reveal new insights and coherencies. Its value is in foregrounding ethical consequences of designing that in turn enable political change or the possibility to create anew in a newly revealed world. We see an expanded practice of design research in which speculative reasoning complements scientific reasoning. This understanding helps to describe practices within traditional research approaches left unexplained by scientific reasoning and gives definition to broad and diverse practices and research that are not squarely based on empiricism or problem-solving.

In defining design research more expansively, our first aim was to redefine speculation not as a genre in the field (i.e., speculative design), a separate and peripheral way of contributing to research, but as a form of reasoning that is utilized widely and diversely in design research. Our framework of leaps of imagination, diverse epistemologies, ethical reflexivity, and experiential alternatives hopefully demonstrated the reach and expansiveness of speculative reasoning in clear terms. Our second aim was to show how characteristics of speculation are pervasive throughout design research. All this to say, that in different combinations, speculative and scientific reasoning co-mingle in the way research is conducted. This interweaving of multiple ways to reason offers a more expansive understanding of reasoning in design research. In this chapter, we turn to the practice of speculative reasoning through a discussion of strategies and methods and we envision the impact on design research when taking the importance of speculation seriously. For convenience, we summarize the aspects of our framework for speculation in Table 5.1. We

© The Author(s), under exclusive license to Springer Nature Switzerland AG 2025   71
R. Wakkary and D. Oogjes, *The Importance of Speculation in Design Research*, Synthesis
Lectures on Human-Centered Informatics, https://doi.org/10.1007/978-3-031-67095-4_5

Table 5.1 Summary of the speculation framework

Leaps of imagination	Diverse epistemologies	Ethical reflexivity	Experiential alternatives
Speculation mobilizes creativity in design through leaps of imagination to experiment with the effects, fosters new ways of thinking, new commitments, and what matters	Speculation explores diverse epistemologies to encourage plurality; is critical of dominant knowledge and seeks the potential of alternatives	Speculation creates space for ethical reflexivity within design by analyzing, demonstrating, or assessing limits and consequences	Speculation is materialized as experiential alternatives that methodologically enable leaps of imagination, epistemological diversity, and ethical reflexivity

Table 5.2 Methods for experiential alternatives

Para-functional things	Things that work otherwise	Materialized stories
The making of technological artifacts or technological referents that function figuratively or metaphorically to make an alternative world experiential	Artifacts and systems that materially function in atypical circumstances following a rationale that is outside current norms	Reenactments or the making of artifacts to create stories that critically investigate dominant assumptions and common knowledge
Para-functional things expand the role of technologies beyond functionality to support speculations that resist norms and subvert assumptions	Things that work otherwise counter the dictates of the role of technologies to reveal unseen or new conditions of our worlds	Materialized stories counter histories and assumptions as settled and singular to reveal obscured, erased, or ignored ways of seeing the world to change the present, past, and future

Related methods we detail in this chapter include Worldbuilding (5.1.6), Counterstories (5.1.7), Counterfactual (5.1.8), and Empirical Subjectivity (5.1.9)

also summarize the methods for experiential alternatives we detailed in earlier chapters and related methods we will elaborate on in this chapter in Table 5.2.

In the book, we used temporal markers of past, present, and future as a rhetorical structure to organize and scaffold our arguments. As a matter of rhetoric and scaffolding, these distinctions are not essential in understanding speculation. In fact, we hope that the reader noticed when we chose examples that undermined the rigidity of this structure, showing speculation to be fluid and seeing temporalities as interdependent. Marie Louise Juul Søndergaard and colleagues describe the work of fabulation as "to trouble temporalities" (Søndergaard et al. 2023, 1693), this overlaps considerably with our own understanding of speculation. Speculation like fabulation's larger commitment is to propositional worlds or "world-making otherwise" (Søndergaard et al. 2023, 1696) rather than the future or the past (Rosner 2023). At this juncture, we remove the scaffolding and move past the rhetorical structure to address speculation within its own inherent fluidity–temporal and otherwise.

In this remaining chapter, we offer guidance on how to put speculative reasoning into practice by describing strategies and methods, followed by reflecting on what design research could be if speculation is seen to be important and central.

5.1 Strategies and Methods for Speculation

The book focused on examples of speculation, the outcomes, and results of various design research practices. These helped to trace the contours and porous outlines of what can be described as speculative reasoning by discussing examples that exemplified the characteristics of our framework. We hope this way of showing by examples encourages further speculative work. Yet we also want to offer guidance on the process or how to use speculative reasoning. To that end, we discuss strategies and methods that we saw in common use among the works discussed in the chapters.

5.1.1 Strategies

Strategies for speculative reasoning function similarly to other strategies as an approach with an overarching aim that best leverages a set of commitments. Discovery and innovation are common strategies in design research based on scientific reasoning. Each leverages the commitments to empirical data, analysis and testing that prove a hypothesis, solve a problem, or address an issue empirically. The aim of speculative reasoning is to enable change or the possibility to create differently in a newly revealed world by focusing on ethical consequences and implications. And so, strategies are unsurprisingly reflective and critical of dominant knowledge and modes. We see three strategies emerging from our discussions that meet this aim: *subversion*, *countering*, and *estrangement*. We will detail each strategy below but for convenience we summarize each in Table 5.3. It's important to note that the strategies are not mutually exclusive, for example in Table 5.5, we compiled the projects and how several examples combine strategies. Relatedly, the experiential alternatives created from each strategy show a wide range of possible outcomes.

When putting speculative reasoning into practice, selecting a strategy is where one typically starts. If the aim is to refute, to offer a clear other possibility to prevailing wisdom then a strategy of countering makes sense. If the desire is to undermine and resist dominant modes of thinking and practices, then a strategy of subversion makes sense. If the aim is to make the familiar unfamiliar to open it to inquiry and new assumptions, then a strategy of estrangement makes sense.

Table 5.3 Strategies

Countering	Subversion	Estrangement
Refuting an idea or a situation with another idea or action. The strategy can include direct opposition, acting to nullify or offer a what-if alternative to the opposed idea. It is an active oppositional stance to dominant norms and assumptions. In speculation, the strategy creates counter-realities in which a different logic is at work or an opposing narrative is offered to show alternatives or to address unjust pasts and the present	The undermining of the power and authority of an established system or mode of thinking. It is often accompanied by irony, which is understood to be an expression that signifies the opposite meaning of what is expressed for humor or emphasis. Together they form an important strategy to realize resistance through design	Making the familiar seem strange by creating discontinuity or distancing that opens it to critical inquiry. The aim is to make an audience aware that normal assumptions do not hold, and the world should not be taken at face value—but rather seen as having the potential to be different

5.1.2   Countering

Countering is to refute an idea or a situation with another idea or action. The strategy can include direct opposition, acting to nullify or offer a what-if alternative to the opposed idea. It is an active oppositional stance to dominant norms and assumptions yet open about what form the stance should take. Given this, it is a common strategy in speculation. Specifically, the strategy creates counter-realities in which a different logic is at work, or an opposing narrative is offered to show alternatives or to address injustices. Several examples throughout our chapters use countering to contest common experiences and underlying assumptions (see Table 5.5). For example, Bikes and Bloomers (Fig. 4.1) utilizes countering by refuting current-day assumptions of innovative technologies and design by offering the "what-if" alternative of fabricating new, subversive fashion patterns of the Victorian era (Jungnickel 2018). Similarly, Pigeon Blog (da Costa, n.d.) (Fig. 3.1) counters scientific and animal rights assumptions about relations with nonhuman species by enacting practices that are neither instrumental nor exploitative but treat the pigeons as co-producers of knowledge and art.

More broadly, we can see aspects of countering in visionary research like Weiser's Ubicomp program that resisted the trend toward the "dramatic machine" in computing (Weiser 1991). Similarly, countering is evident in the theoretical analyses in Jackson's *broken world thinking* that critiques notions of endless development and novelty (Jackson 2014), countering norms of progress to seek innovations in the breakdowns of technology, the creative practice in the labor of maintenance and repair.

5.1.3 Subversion

Subversion is the undermining of the power and authority of an established system or mode of thinking. It is an important strategy to realize resistance through design. In speculation, irony and subversion undercut typical means of expression, actions or commonly held assumptions and underscores this by creatively working within the opposition. For example, it can be accompanied by irony to leverage humor to enrich the form of opposing ideas. The aim of subversion is to use oppositional tension to create new and propositional space that is rich and multidimensional rather than a simple inversion or straightforward opposition. Subversion is common in Italian Radical Design that expressly challenged the political authorities and the dictates of industrial capitalism and its withering of design. The *Superarchitettura* manifesto adopted a clear stance of subversion. It subverted the logics of consumer capitalism by hyper-inflating the claims of the exhibition, an exhibition that included no functional design objects. Another good example is Bray and Harrington's *Building Utopia* design cards subvert the power of systemic racism to centralize the marginalized to "reimagine potential futures" (Fig. 2.3). They draw on Afrofuturism that can be ironic but above all subverts the implicit assumption of ongoing narratives of whiteness into the future.

Traces of subversion can be seen in other works we cited that do not foreground speculative reasoning yet rely in part on speculation. Critical literature reviews like those we mentioned, expose gaps in the literature to subvert behavior change toward politics of technology design in sustainability research (DiSalvo et al. 2010). Similarly, design games in participatory design undermine technocentric approaches to democratize expertise in designing workplaces (Ehn and Kyng 1992).

5.1.4 Estrangement

Estrangement is to make the familiar seem strange to open it to critical inquiry. The strategy distances us from our typical perceptions and assumptions of our daily lives with the aim to make an audience critically aware that the world should not be taken at face value–but rather seen as holding the potential to be different. Often, the strategy relies on creating a discontinuity within normal practices or situations. This in turn makes the most accepted practices and situations open to speculation. For example, mundane things like bowls can through small, even trivial interventions like tilting, can be seen differently, opening the door for new interpretations of how we relate to technologies as with the Tilting Bowl (Wakkary et al. 2015). Similarly, the interventions that result from estrangement, like wearing a mask to view a neighborhood as a nonhuman animal, can dislodge us from our typical perspectives (Clarke 2020). Estrangement is connected to a longer history of similar strategies. For example, the estrangement of the Plant Hotels (Wu and Koskinen 2022) to make strange mundane routines to explore new social possibilities adapts a longstanding ethnographic method of *breaching* (Garfinkel 1964).

In summary, the three strategies we described are our interpretations of commonalities across the speculative examples we discussed in the book. They each position themselves in relation to dominant thinking as a speculative approach to create propositional space and the potential within for new possibilities. As discussed, the methods can be combined in various ways as different aspects of the experiential alternative emphasize or utilize different strategies. We next describe the methods that mobilize these strategies.

5.1.5 Methods

In this section, we describe additional methods we found in common across the examples of the book, in addition to *para-functional things, things that work otherwise*, and *materialized stories* we discussed in prior chapters. If strategies set the aim for doing speculative research, methods are the means and processes by which the strategies are carried out. The aim of the methods is to guide the creation of the experiential alternatives, what form will it take and what processes will it rely upon in its making. Unsurprisingly, the methods differ from design research that rely less upon speculative reasoning. The overarching aims are to support the countering of dominant beliefs, foster the conditions for provisional thought that can lead to alternatives or propositions to be realized, and rely upon distinct ways of knowing the world. The additional methods we detail here are *worldbuilding, counterstories, counterfactual,* and *empirical subjectivity*. Like strategies, we will detail each below but for convenience, we summarize the methods in Table 5.4. We compiled all the projects we discussed in this book in Table 5.5. The table includes the methods and strategies that we believe are in operation with each example. These are our interpretations. Nevertheless, the table shows that while the methods are not separate from the strategies, there are multiple ways to combine different methods with different strategies. This is a powerful feature as it shows the capacity and range of speculative research in design.

Table 5.4 Additional Methods

Worldbuilding	Counterstories	Counterfactual	Empirical subjectivity
Creating a fictional world that makes a convincing illusion, an experiential alternative from which to consider other futures	Telling untold stories through narrative structures to counter the dominance of the majoritarian stories	Asking "what-if" questions that counter reality in any number of ways	Prioritizing unique viewpoints and perspectives over that of generalizable majorities through first-person and third-person approaches

Table 5.5 Mapping the projects to methods and strategies

Project	Methods	Strategy
Bespoke probes (Wallace et al. 2013)	Counterstories, empirical subjectivity, things that work otherwise	Countering
Menstruation machine (Ozaki 2010)	Worldbuilding, counterfactual, para-functional things	Countering
PigeonBlog (da Costa n.d.)	Empirical subjectivity, counterstories, things that work otherwise	Countering
Audio tooth implant (Auger and Loizeau 2001)	Counterfactual, worldbuilding, para-functional things	Countering/subversion
Biographical prototypes (Bennett et al. 2019)	Counterstories, things that work otherwise	Countering
Weaving stories (Oogjes and Wakkary 2022)	Counterstories, counterfactual, things that work otherwise	Countering/estrangement
Bikes and bloomer (Jungnickel 2018)I	Counterstories, counterfactual, materialized stories	Countering
Before yesterday We Could Fly: an Afrofuturist period room (Alteveer et al. 2022)	Counterstories, materialized stories	Countering/subversion
Making core memory (Rosner et al. 2018)	Counterstories, materialized stories	Countering
Counterfactual actions (Forlano and Halpern 2023)	Counterfactual, materialized stories	Countering
Plant Hotels (Wu and Koskinen 2022)	Counterfactual, worldbuilding, things that work otherwise	Countering/estrangement
Tilting Bowl	Counterfactual, empirical subjectivity, things that work otherwise	Countering/estrangement
Ministry of Multispecies Communication (Clarke 2020)	Worldbuilding, counterfactual, empirical subjectivity, materialized stories	Estrangement
Alternative Metaphors from Dark Clouds, Io$#!+, and 🜚[Crystal Ball Emoji] (Pierce and DiSalvo 2017)	Counterfactual, para-functional things	Countering
Placebo Project (Dunne and Raby 2001)	Counterfactual, para-functional things	Estrangement
Watching myself watch birds	Empirical subjectivity, things that work otherwise	Estrangement
Superarchitettura Manifest (Superstudio and Archizoom)	Counterfactual	Subversion

(continued)

Table 5.5 (continued)

Project	Methods	Strategy
Urboeffimero (UFO)	Counterfactual, para-functional things	Subversion
Building Utopia Cards (Bray and Harrington 2021)	Worldbuilding, materialized stories	Subversion
Supersurface: an alternative model for life on Earth (Superstudio)	Worldbuilding, para-functional things	Subversion/countering
RIOT (Fox et al. 2018)	Empirical subjectivity, things that work otherwise	Subversion
Hands Up, Don't Shoot Glove (Chin 2022)	Counterstories, empirical subjectivity, things that work otherwise	Subversion/countering
Decolonial imaginary (Barcham 2022)	Counterstories, materialized stories	Subversion

5.1.6 Worldbuilding

Worldbuilding is a common method in speculation referring to the making of a fictional world or parts of that world experiential. In other words, a world that shows the possibilities that are revealed through speculative reasoning.

Worldbuilding utilizes techniques including the crafting of a story or narrative that surrounds the world, including nuances and aesthetic details. For example, the Menstruation Machine (Chap. 2) is presented as a music video that shows Takashi preparing to go out (doing their make-up, selecting an outfit) and them wearing the machine in the city. The choice of music and other aesthetic details contribute to the building of the story world around Menstruation Machine. The *Menstruation Machine* is often referred to as design fiction in which the design object can be seen as a *diegetic prototypes* (Kirby 2010) whose role is to move the story along (or diegesis). The object itself is another key part of worldbuilding as a method, whether that be a full artifact, prop, or a probe. For example, the use of playing cards in Utopia cards to make experiential or build a different world. In other examples, the form of visualization can be through illustrative drawings, photocollages (such as in Supersurface: An Alternative Model for Life on Earth), creating fictional news clippings such as in Plant Hotel or product launches at fairs such as with Audio Tooth.

In speculation, the making of a fictional world is about creating a convincing illusion, an experiential alternative from which to consider propositional alternatives.

5.1.7   Counterstories

Counterstories tell untold stories through narrative structures to counter the dominance of the majoritarian stories. Dominant stories or narratives are built on a deep interweaving of assumptions that need unpacking and unsettling given their hold on the trajectories of the present that often reproduces harms through racism, sexism, colonialism, technological determinism, and human exceptionalism. This method relates to *counterstorytelling* of critical race theory and anti-racism (Ogbonnaya-Ogburu et al. 2020), or *counterventions* of the research on autism that we discussed countering neurotypical assumptions that underpin much of this research (Williams, Boyd, and Gilbert 2023).

In the speculative examples presented in the chapters, we see different ways in which counterstories offer possibilities for revealing and repairing (see examples in Table 5.5). With *Before Yesterday we Could Fly*, the erased story of Black Seneca Village is not simply told again but the very logic and structure of the storytelling are upended and explodes into a beautiful fracturing of the stories within the stories that materialize as a "period room" (Alteveer et al. 2022). The room does not adhere to what was in the past but simultaneously to what could have been and what is in the present of the stories of Black Seneca Village. With counterstories, the countering of the dominant story is accompanied by the radical restructuring of the form of the narrative itself. Other examples utilize personal stories (such as in Biographical Prototypes (Bennett, Peil, and Rosner 2019)), re-enact stories (such as in Bikes and Bloomers (Jungnickel 2018)) or experiment with different forms of narrative to bring to the fore other agencies. Counterstories create stories—be it through text, illustration, or installations—that oppose dominant narratives to retell stories that are obscured and erased. Counterstories also counter the very structures by which dominant stories and practices are told or enacted by offering expressive logics and multiplicities to the way we can encounter and shape the present.

5.1.8   Counterfactual

Counterfactual is the asking of "what-if" questions that counter reality. Counterfactual extends beyond functionality—it is often used in philosophy and key to many thought experiments such as Einstein's *train and platform* or Ibn Sina's *flying man* examples discussed in the introduction. Common to philosophy is the question of knowing the world differently and how to do that. The counterfactual allows philosophers to reason on an imagined alternative. This directly relates to literature that extensively utilizes counterfactuals in science fiction and magical realism, such as the writer Haruki Murakami, in which possible worlds intersect to counter our actual world. In design, we have argued for the role of *counterfactual artifacts* in an approach we call Material Speculations (Wakkary et al. 2015). Here the countering strategy is to design an artifact that would not normally be designed, given the norms of human-centered design, functionalism, or social norms, yet by virtue of being designed opens a possible world to inquire as a matter of research. For example, the Plant Hotel can be described as a breaching experiment, as the authors do, to consciously create unexpected behaviors as a mode of observation. It can also be seen as counterfactual in that one would not normally expect strangers to care for others' plants without any reciprocity (Wu and Koskinen 2022). The underscoring of the counterfactual approach is to propose the Plant Hotel counterfactual to address the estranged populations of war in North and South Korea. Counterfactual artifacts rely on a careful balance of what might seem absurd (given design norms) yet remain acceptable in that they can be lived with in everyday life. Counterfactual as a speculative method is generative of new possibilities and perspectives.

5.1.9 Empirical Subjectivity

Empirical subjectivity utilizes the observable and experiential every day and includes approaches such as field studies (Fox, Silva, and Rosner 2018) and is related to approaches such as speculative enactments (Elsden et al. 2017) and co-speculation (Wakkary et al. 2017; Desjardins et al. 2019; Wakkary, Oogjes, and Behzad 2022). What is specific about the speculative methods is the prioritizing of unique viewpoints and perspectives over that of generalizable majorities. For example, through first-person point of view of the researcher or designer, such as Heidi Biggs' Birdwatching project (Biggs, Bardzell, and Bardzell 2021). By putting their own reflections and descriptions of events central, these methods make use of the situated, embodied, and personal experiences that offer specific, nuanced, and particular insights that would be glanced over in more universal approaches. The speculative work in Biggs' birdwatching draws from recollections of their dreams, their visceral experiences when birdwatching in the swamp and reflections on notes taken when encountering birds in everyday life. Taken together, these enabled Biggs to create the laugh-track video that richly communicates their experience of abjection. First-person methods often rely on data collection such as field notes, audio recordings, diary entries and photographs. There is a growing range of first-person methods relevant for design research, including approaches such as autobiographical design, design memoirs and first-person ethnography or duo and trio-ethnography that can be used in speculation.

Unique viewpoints are leveraged through the choice of participant or co-speculator—people in particular positions rather than a "neutral" or universal one. For example *Hand's Up, Don't Shoot Glove* (Chin 2022), in drawing from the lived experiences of black people in the United States in designing for the connection between technology and spirituality. In an example of our own, we chose to co-speculate with trained philosophers in our study with the Tilting Bowl (Wakkary et al. 2018), as they are in a unique position through their analytical skills and daily practice to critically think along with us on matters of philosophy of technology.

Empirical subjectivity in speculation foregrounds unique voices and the particularities of their lived experience—including the materials in their everyday environment—using first-person methods such as autoethnography and autobiographical design, as well as third-person methods such as field studies, co-speculation, speculative enactments, and more.

We have described strategies and methods used in the examples of the book. As we hinted at previously, we see these as porous and complementary—one method does not fit with only one strategy and vice versa, and multiple combinations are possible (see Table 5.5). As well, there are notable strategies and methods used in speculative work that we have not named or covered here, including for example emergent (W. Gaver et al. 2022), reflective (Kozubaev et al. 2020), ambiguous (W. W. Gaver, Beaver, and Benford 2003; Sengers and Gaver 2006), poetic (Sharma et al. 2022; Bozic Yams and Aranda Muñoz 2021), and soulful speculation (Halperin and Rosner 2023). Our goal was not to be comprehensive but to give starting points for further speculative work.

5.2 Speculation Is Important to Design Research

In this book, we argued that speculation is a powerful resource for design research. Speculation enables constructing propositional knowledge, countering dominant worldviews and revealing alternative or new forms of knowledge. We positioned speculation as a form of reasoning with incomplete knowledge. In the introduction, we presented our speculation framework of leaps of imagination, diverse epistemologies, ethical reflections, and experiential alternatives. As a rhetorical structure, we created temporal framings of the continuous future, continuous present, and continuous past to structure our argument for speculative reasoning. This temporal framing allowed us to show the reach and expansiveness of speculation to include current and past practices as well as the future. In these chapters, we used our framework to detail diverse examples of speculative reasoning in design research to show the prevalence and depth of speculation in our fields. We also demonstrated how aspects of speculation can be traced to existing practices in design research that would typically not be seen as speculative though they clearly employ speculative reasoning co-mingling with scientific reasoning. As was our aim, we hoped to show that speculation is not secluded within a genre of speculative design, but rather is central to design research.

In this final chapter, we removed the rhetorical scaffolding of the past, present, and future to see speculative reasoning as fluid and multidimensional. We then discussed strategies and methods that we saw in common across the work in the chapters to offer guidance on how to use speculative reasoning, including: subversion, countering and defamiliarization as strategies; and worldbuilding, counterstories, counterfactual, empirical subjectivity, para-functional things, things that work otherwise, and materialized stories as methods.

We are motivated by the belief that speculation will be increasingly important to design research given the pace of technological developments, economic uncertainty, global health crises, and climate emergencies. We offer speculation not as a new novelty but rather to tap into the long history of speculation to address the ever-present instabilities, uncertainties, and incomplete knowledge of our world, for example Lord Bryon's poem in response to abnormal climate change in 1816 and Virginia Woolf's uncertainty living in the midst of the First World War in her writing in 1915. The value of speculation is not to affirm these uncertainties or to find stabilities and solutions—rather, its power lies in creating conditions to think and act with the incomplete and unknown propositionally, creating a space of experimentation for the potential consequences, possible outcomes, and felt experiences.

Looking forward, we end by turning to the question: *what would design research look like if it took seriously the importance of speculation?*

This is a leading question. The claim of this book is that speculation be seen as a central mode of thinking in design research alongside other modes of reasoning, like scientific reasoning. The importance of the claim is the increasing necessity for design to address current and ongoing instabilities and uncertainties like the climate crisis, decolonization, or racism. What is particular about these challenges is how design's successes have contributed to them. The unrelenting pervasiveness of designing technologies, especially digital technologies, means that now more than ever, design shapes everyday realities, human behaviors, social structures, and our broader world such that design is enmeshed within these concerns, implicated if not complicit.

The increasing entanglement of design research in everyday realities makes precarity and complexity the norm rather than the exception. This calls for a generosity in rethinking design research, a greater plurality in ways of researching. Challenges like climate crisis, decolonization, or racism do not lend themselves to reductionism and causality. Researching innovative technologies is no guarantee of progress, and to seek progress without asking for whom and at what cost to others (including those that are not human) is insufficient. Scientific reasoning operates best within certitudes of knowledge—high degrees of confidence in what one knows. However, such clarity is often elusive in the indeterminacy or sheer complexity of the challenges we face. Speculative reasoning adds the potential to research with what is not known rather than exclusively with what is known. Our argument is a declaration and demonstration that speculative reasoning not only *holds the potential* to be central to design research but *should be called upon* to be central to design research.

And so, what will design research look like if we take speculation seriously? In part, through the examples and discussions in this book, we have been sketching a version of design research that holds speculation central to its practices. It is a generous endeavor that further solidifies the importance and entanglement of designing technology with matters of politics, race, gender, inclusion, multispecies, feminism, democracy, indigeneity, or decolonization. Not as matters to solve in the problem-solving sense but as matters to productively engage within, to navigate the differences, contradictions, and opportunities by working through implications and consequences through propositional knowledge. A more speculative design research will more comfortably embrace plurality. This could mean exploring propositional spaces at their limits or traversing into what was previously seen as off-limits. Researching in ways that are not conditional on working at the center, exclusively building on what is known but venturing to the boundaries and past them to explore building on what is not known or hard to know. Plurality would also come in the form of embracing different ways of knowing in various combinations, as we explored in our examples of diverse epistemologies. This more generous design research would be open to a multiplicity of voices and actors beyond the current molds of the scientist, technologist, or designer—and hopefully more accessible to those that are not empowered by the disciplinary traditions of science, technology, and design.

Despite our optimism, speculation is by no means a panacea. It is as subject to folly, failure, misuse, and vanity as any other approach. It is subject to good and bad speculation and all that goes in between. We offer it generously as an addition not only to the current understanding of the genre of speculative design but also to design research. It should not be valorized above other approaches; it is a unique addition with its own limitations.

Interestingly, the uniqueness of speculation is to bring to the fore the need to work with our limitations, both the dangers and joys these might afford. For Margaret Atwood, literature is bringing forth the intermingling of our fears and joys. Similarly, we can say that speculation is "an uttering or outering of the human imagination. It lets the shadowy forms of thought and feeling—Heaven, Hell, monsters, angels, and all—out into the light, where we can take a good look at them and perhaps come to a better understanding of who we are and what we want, and what the limits to those wants may be" (Atwood 2023, 13).

References

Agre, Philip. 1997. *Computation and Human Experience.* Cambridge University Press.

Ahmed, Sara. 2006. *Queer Phenomenology: Orientations, Objects, Others.* Durham: Duke University Press.

Aldiss, Brian. 1995. *The Detached Retina: Aspects of SF and Fantasy.* Syracuse University Press.

Alteveer, Ian, Hannah Beachler, Michael D. Commander, and John Jennings. 2022. "Before Yesterday We Could Fly: An Afrofuturist Period Room." *The Metropolitan Museum of Art Bulletin*, MetPublications - The Metropolitan Museum of Art, 79 (3): 52 pages.

Andersen, Kristina, and Ron Wakkary. 2019. "The Magic Machine Workshops: Making Personal Design Knowledge." In *Proceedings of the 2019 CHI Conference on Human Factors in Computing Systems*, 1–13. CHI '19. New York, NY, USA: Association for Computing Machinery. https://doi.org/10.1145/3290605.3300342.

Ansari, Ahmed. 2019a. "Decolonizing Design through the Perspectives of Cosmological Others: Arguing for an Ontological Turn in Design Research and Practice." *XRDS: Crossroads, The ACM Magazine for Students* 26 (2): 16–19. https://doi.org/10.1145/3368048.

Ansari, Ahmed. 2019b. "The Politics of Critical Design: Jamer Hunt vs. Ahmed Ansari." *Medium* (blog). February 6, 2019. https://aansari86.medium.com/design-must-fi-ll-current-human-needs-before-imagining-new-futures-7a9b10815342.

Arets, Daniëlle Joseph Antonius Maria. 2024. "Save the Debate: Through Adversarial Design." Phd Thesis 1 (Research TU/e / Graduation TU/e), Eindhoven: Eindhoven University of Technology.

Atwood, Margaret. 2023. *Burning Questions: Essays and Occasional Pieces, 2004 to 2022.* Vintage.

Auger, James. 2010. "Alternative Presents and Speculative Futures: Designing Fictions through the Extrapolation and Evasion of Product Lineages." *Negotiating Futures – Design Fiction.* 6 (October): 42–57.

Austin, John L. 1962. *How to Do Things with Words.* Oxford, UK: Oxford University Press.

Balen, Artür van. 2018. "Floating Utopias: Urboeffimiri/Skala 1:1 - an Interview by Artür van Balen with Lapo Binazzi." Blog. Floating Utopias. March 2018. https://floatingutopias.org/stories/lapo-binazzi-artur-van-balen/.

Balka, Ellen, and Ina Wagner. 2021. "A Historical View of Studies of Women's Work." *Computer Supported Cooperative Work (CSCW)* 30 (2): 251–305. https://doi.org/10.1007/s10606-020-09387-9.

Banks, Richard, David Kirk, and Abigail Sellen. 2012. "A Design Perspective on Three Technology Heirlooms." *Human–Computer Interaction* 27 (1–2): 63–91. https://doi.org/10.1080/07370024.2012.656042.

R. Wakkary and D. Oogjes, *The Importance of Speculation in Design Research*, Synthesis Lectures on Human-Centered Informatics, https://doi.org/10.1007/978-3-031-67095-4

Barcham, Manuhuia. 2022. "Weaving Together a Decolonial Imaginary Through Design for Effective River Management: Pluriversal Ontological Design in Practice." *Design Issues* 38 (1): 5–16. https://doi.org/10.1162/desi_a_00666.

Bardzell, Jeffrey, and Shaowen Bardzell. 2013. "What Is 'Critical' About Critical Design?" In *Proceedings of the SIGCHI Conference on Human Factors in Computing Systems*, 3297–3306. CHI '13. New York, NY, USA: ACM. https://doi.org/10.1145/2470654.2466451.

Bardzell, Jeffrey, Shaowen Bardzell, and Lone Koefoed Hansen. 2015. "Immodest Proposals: Research Through Design and Knowledge." In *Proceedings of the 33rd Annual ACM Conference on Human Factors in Computing Systems*, 2093–2102. CHI '15. New York, NY, USA: Association for Computing Machinery. https://doi.org/10.1145/2702123.2702400.

Bardzell, Shaowen. 2010. "Feminist HCI: Taking Stock and Outlining an Agenda for Design." In *Proceedings of the SIGCHI Conference on Human Factors in Computing Systems*, 1301–10. CHI '10. New York, NY, USA: ACM. https://doi.org/10.1145/1753326.1753521.

Baytaş, Mehmet Aydin, Aykut Coşkun, Asim Evren Yantaç, and Morten Fjeld. 2018. "Towards Materials for Computational Heirlooms: Blockchains and Wristwatches." In *Proceedings of the 2018 Designing Interactive Systems Conference*, 703–17. DIS '18. New York, NY, USA: Association for Computing Machinery. https://doi.org/10.1145/3196709.3196778.

Bell, Genevieve, Mark Blythe, and Phoebe Sengers. 2005. "Making by Making Strange: Defamiliarization and the Design of Domestic Technologies." *ACM Trans. Comput.-Hum. Interact.* 12 (2): 149–73. https://doi.org/10.1145/1067860.1067862.

Bell, Genevieve, and Paul Dourish. 2007. "Back to the Shed: Gendered Visions of Technology and Domesticity." *Personal and Ubiquitous Computing* 11 (5): 373–81. https://doi.org/10.1007/s00779-006-0073-8.

Bennett, Cynthia L., Burren Peil, and Daniela K. Rosner. 2019. "Biographical Prototypes: Reimagining Recognition and Disability in Design." In *Proceedings of the 2019 on Designing Interactive Systems Conference*, 35–47. DIS '19. New York, NY, USA: Association for Computing Machinery. https://doi.org/10.1145/3322276.3322376.

Biggs, Heidi R., Jeffrey Bardzell, and Shaowen Bardzell. 2021. "Watching Myself Watching Birds: Abjection, Ecological Thinking, and Posthuman Design." In *Proceedings of the 2021 CHI Conference on Human Factors in Computing Systems*, 1–16. CHI '21. New York, NY, USA: Association for Computing Machinery. https://doi.org/10.1145/3411764.3445329.

Bleecker, Julian. 2009. "Design Fiction: A Short Essay on Design, Science, Fact and Fiction." Near Future Laboratory. http://drbfw5wfjlxon.cloudfront.net/writing/DesignFiction_WebEdition.pdf.

Blevis, Eli. 2007. "Sustainable Interaction Design: Invention & Disposal, Renewal & Reuse." In *Proceedings of the SIGCHI Conference on Human Factors in Computing Systems*, 503–12. CHI '07. New York, NY, USA: ACM. https://doi.org/10.1145/1240624.1240705.

Bødker, Susanne. 2006. "When Second Wave HCI Meets Third Wave Challenges." In *Proceedings of the 4th Nordic Conference on Human-Computer Interaction: Changing Roles*, edited by Anders Mørch, Konrad Morgan, Tone Bratteteig, Gautam Ghosh, and Dag Svanaes, 1–8. NordiCHI '06. Oslo, Norway: Association for Computing Machinery. https://doi.org/10.1145/1182475.1182476.

Bozic Yams, Nina, and Álvaro Aranda Muñoz. 2021. "Poetics of Future Work: Blending Speculative Design with Artistic Methodology." In *Extended Abstracts of the 2021 CHI Conference on Human Factors in Computing Systems*. CHI EA '21. New York, NY, USA: Association for Computing Machinery. https://doi.org/10.1145/3411763.3443451.

Braidotti, Rosi. 2013. *The Posthuman*. Cambridge, UK ; Malden, MA, USA: Polity.

Bray, Kirsten E, Christina Harrington, Andrea G Parker, N'Deye Diakhate, and Jennifer Roberts. 2022. "Radical Futures: Supporting Community-Led Design Engagements through an Afrofuturist Speculative Design Toolkit." In *Proceedings of the 2022 CHI Conference on Human Factors in Computing Systems*. CHI '22. New York, NY, USA: Association for Computing Machinery. https://doi.org/10.1145/3491102.3501945.

Bray, Kirsten, and Christina Harrington. 2021. "Speculative Blackness: Considering Afrofuturism in the Creation of Inclusive Speculative Design Probes." In *Designing Interactive Systems Conference 2021*, 1793–1806. DIS '21. New York, NY, USA: Association for Computing Machinery. https://doi.org/10.1145/3461778.3462002.

Burton, Michael, and Michiko Nitta. 2011. *Republic of Salvation*. Insatallation. https://www.burton nitta.co.uk/RepublicOfSalivation.html.

Carroll, J. M., and W. A. Kellogg. 1989. "Artifact As Theory-Nexus: Hermeneutics Meets Theory-Based Design." In *Proceedings of the SIGCHI Conference on Human Factors in Computing Systems*, 7–14. CHI '89. New York, NY, USA: ACM. https://doi.org/10.1145/67449.67452.

Chin, Elizabeth. 2022. "Speculating Spiritual Technologies." *Interactions* 29 (4): 60–61. https://doi.org/10.1145/3542699.

Clarke, Rachel E. 2020. "Ministry of Multispecies Communications." In *Companion Publication of the 2020 ACM Designing Interactive Systems Conference*, 441–44. DIS' 20 Companion. New York, NY, USA: Association for Computing Machinery. https://doi.org/10.1145/3393914.3395845.

Clarke, Rachel, Sara Heitlinger, Ann Light, Laura Forlano, Marcus Foth, and Carl DiSalvo. 2019. "More-than-Human Participation: Design for Sustainable Smart City Futures." *Interactions* 26 (3): 60–63. https://doi.org/10.1145/3319075.

Cohen, Revital, and Tuur Van Balen. 2008. "LIFE SUPPORT - REVITAL COHEN & TUUR VAN BALEN." Life Support. 2008. https://www.cohenvanbalen.com/work/life-support.

Costa, Beatriz da. n.d. "PigeonBlog." Accessed August 2, 2023. https://isea-archives.siggraph.org/art-events/beatriz-da-costa-pigeonblog/.

Crabtree, Andy. 2004. "Design in the Absence of Practice: Breaching Experiments." In *Proceedings of the 5th Conference on Designing Interactive Systems: Processes, Practices, Methods, and Techniques*, 59–68. DIS '04. New York, NY, USA: Association for Computing Machinery. https://doi.org/10.1145/1013115.1013125.

Cryer, Beryl Mildred. 2008. *Two Houses Half-Buried in Sand: Oral Traditions of the Hul'q'umi'num' Coast Salish of Kuper Island and Vancouver Island*. Edited by Chris Arnett. Reprint edition. Vancouver: Talonbooks.

Debaise, Didier, and Isabelle Stengers. 2017. "The Insistence of the Possible." *Parse*, no. 6 (Autumn): 12–19.

Desjardins, Audrey, Cayla Key, Heidi R. Biggs, and Kelsey Aschenbeck. 2019. "Bespoke Booklets: A Method for Situated Co-Speculation." In *Proceedings of the 2019 on Designing Interactive Systems Conference*, 697–709. DIS '19. New York, NY, USA: Association for Computing Machinery. https://doi.org/10.1145/3322276.3322311.

Desjardins, Audrey, Ron Wakkary, and William Odom. 2015. "Investigating Genres and Perspectives in HCI Research on the Home." In *Proceedings of the 33rd Annual ACM Conference on Human Factors in Computing Systems*, 3073–82. CHI '15. New York, NY, USA: ACM. https://doi.org/10.1145/2702123.2702540.

Despret, Vinciane. 2021. *Our Grateful Dead: Stories of Those Left Behind*. Translated by Stephen Muecke. Minneapolis: Univ Of Minnesota Press.

Didero, Maria Cristina, Evan Snyderman, Deyan Sudjic, and Catharine Rossi. 2017. *SuperDesign: Italian Radical Design 1965-75*. New York, N.Y: The Monacelli Press.

DiSalvo, Carl. 2012. *Adversarial Design*. The MIT Press. https://doi.org/10.7551/mitpress/8732.001.0001.

DiSalvo, Carl, Phoebe Sengers, and Hrönn Brynjarsdóttir. 2010. "Mapping the Landscape of Sustainable HCI." In *Proceedings of the SIGCHI Conference on Human Factors in Computing*

Systems, 1975–84. CHI '10. New York, NY, USA: ACM. https://doi.org/10.1145/1753326.175 3625.

Dunne, Anthony. 2008. *Hertzian Tales: Electronic Products, Aesthetic Experience, and Critical Design.* Cambridge, Mass.: MIT Press.

Dunne, Anthony, and Fiona Raby. 2001. *Design Noir: The Secret Life of Electronic Objects.* Springer. http://books.google.ca/books?hl=en&lr=&id=_49YTKJ16l4C&oi=fnd&pg=PA6&dq= design+noir&ots=5HmalnNO9U&sig=-E2yuIbwixhJf7JBv9v9f9Nxhhc.

Dunne, Anthony, and Fiona Raby. 2013. *Speculative Everything: Design, Fiction, and Social Dreaming.* Cambridge, Massachusetts ; London: The MIT Press.

Ehn, Pelle. 1988. "Work-Oriented Design of Computer Artifacts." https://urn.kb.se/resolve?urn=urn: nbn:se:umu:diva-62913.

Ehn, Pelle, and Morten Kyng. 1992. "Cardboard Computers: Mocking-It-up or Hands-on the Future." In *Design at Work: Cooperative Design of Computer Systems*, 169–96. USA: L. Erlbaum Associates Inc.

Elsden, Chris, David Chatting, Abigail C. Durrant, Andrew Garbett, Bettina Nissen, John Vines, and David S. Kirk. 2017. "On Speculative Enactments." In *Proceedings of the 2017 CHI Conference on Human Factors in Computing Systems*, 5386–99. CHI '17. New York, NY, USA: ACM. https:// doi.org/10.1145/3025453.3025503.

Elsden, Chris, and David S. Kirk. 2014. "A Quantified Past: Remembering with Personal Informatics." In *Proceedings of the 2014 Companion Publication on Designing Interactive Systems*, 45–48. DIS Companion '14. New York, NY, USA: Association for Computing Machinery. https://doi. org/10.1145/2598784.2602778.

Fastag, Eduardo, Joseph Varon, and George Sternbach. 2013. "Richard Lower: The Origins of Blood Transfusion." *The Journal of Emergency Medicine* 44 (6): 1146–50. https://doi.org/10.1016/j.jem ermed.2012.12.015.

Fedosov, Anton, William Odom, Marc Langheinrich, and Ron Wakkary. 2018. "Roaming Objects: Encoding Digital Histories of Use into Shared Objects and Tools." In *Proceedings of the 2018 Designing Interactive Systems Conference*, 1141–53. DIS '18. New York, NY, USA: Association for Computing Machinery. https://doi.org/10.1145/3196709.3196722.

Fernaeus, Ylva, Martin Jonsson, and Jakob Tholander. 2012. "Revisiting the Jacquard Loom: Threads of History and Current Patterns in HCI." In *Proceedings of the SIGCHI Conference on Human Factors in Computing Systems*, 1593–1602. CHI '12. New York, NY, USA: ACM. https:// doi.org/10.1145/2207676.2208280.

Forlano, Laura E., and Megan K. Halpern. 2023. "Speculative Histories, Just Futures: From Counterfactual Artifacts to Counterfactual Actions." *ACM Transactions on Computer-Human Interaction* 30 (2): 1–37. https://doi.org/10.1145/3577212.

Forlizzi, Jodi, John Zimmerman, Paul Hekkert, and Ilpo Koskinen. 2018. "Let's Get Divorced: Constructing Knowledge Outcomes for Critical Design and Constructive Design Research." In *Proceedings of the 2018 ACM Conference Companion Publication on Designing Interactive Systems*, 395–97. DIS '18 Companion. New York, NY, USA: Association for Computing Machinery. https://doi.org/10.1145/3197391.3197395.

Fox, Sarah E., Rafael M.L. Silva, and Daniela K. Rosner. 2018. "Beyond the Prototype: Maintenance, Collective Responsibility, and Public IoT." In *Proceedings of the 2018 Designing Interactive Systems Conference*, 21–32. DIS '18. New York, NY, USA: ACM. https://doi.org/10. 1145/3196709.3196710.

Frassinelli, Piero, Alessandro Magris, Roberto Magris, Adolfo Natalani, Alessandro Poli, and Cristiano Toraldo di Francia. 1972. "Superstudio." In *Italy: The New Domestic Landscape*, edited by Emilio Ambasz, 240–50. New York: Museum of Modern Art.

Frauenberger, Christopher. 2019. "Entanglement HCI The Next Wave?" *ACM Transactions on Computer-Human Interaction* 27 (1): 2:1–2:27. https://doi.org/10.1145/3364998.

Frens, J.W. 2006. "Designing for Rich Interaction : Integrating Form, Interaction, and Function: Conference; 3rd Symposium of Design Research; 2006-11-17; 2006-11-18." *Drawing New Territories; 3rd Symposium of Design Research, Swiss Design Network, Switzerland*, 91–106.

Gadamer, Hans-Georg. 1976. *Philosophical Hermeneutics*. Translated by David E. Linge. University of California Press.

Garfinkel, Harold. 1964. "Studies of the Routine Grounds of Everyday Activities." *Social Problems* 11 (3): 225–50. https://doi.org/10.2307/798722.

Gaver, William, Peter Gall Krogh, Andy Boucher, and David Chatting. 2022. "Emergence as a Feature of Practice-Based Design Research." In *Designing Interactive Systems Conference*, 517–26. DIS '22. New York, NY, USA: Association for Computing Machinery. https://doi.org/10.1145/3532106.3533524.

Gaver, William W., Jacob Beaver, and Steve Benford. 2003. "Ambiguity as a Resource for Design." In, 233. ACM Press. https://doi.org/10.1145/642611.642653.

Gulotta, Rebecca, William Odom, Jodi Forlizzi, and Haakon Faste. 2013. "Digital Artifacts as Legacy: Exploring the Lifespan and Value of Digital Data." In *Proceedings of the SIGCHI Conference on Human Factors in Computing Systems*, 1813–22. CHI '13. New York, NY, USA: Association for Computing Machinery. https://doi.org/10.1145/2470654.2466240.

Gulotta, Rebecca, Alex Sciuto, Aisling Kelliher, and Jodi Forlizzi. 2015. "Curatorial Agents: How Systems Shape Our Understanding of Personal and Familial Digital Information." In *Proceedings of the 33rd Annual ACM Conference on Human Factors in Computing Systems*, 3453–62. CHI '15. New York, NY, USA: Association for Computing Machinery. https://doi.org/10.1145/2702123.2702297.

Halperin, Brett A., and Daniela K. Rosner. 2023. "Miracle Machine in the Making: Soulful Speculation with Kabbalah." In *Proceedings of the 2023 ACM Designing Interactive Systems Conference*, 1740–56. DIS '23. New York, NY, USA: Association for Computing Machinery. https://doi.org/10.1145/3563657.3595990.

Hamilton, Virginia. 1993. *The People Could Fly: American Black Folktales*. Illustrated edition. New York: Knopf Books for Young Readers.

Haraway, Donna. 2016. *Staying with the Trouble: Making Kin in the Chthulucene*. First Edition edition. Durham: Duke University Press Books.

Harrison, Steve, Deborah Tatar, and Phoebe Sengers. 2007. "The Three Paradigms of HCI." *Alt. Chi.* http://people.cs.vt.edu/~srh/Downloads/HCIJournalTheThreeParadigmsofHCI.pdf.

Hartman, Saidiya. 2008. "Venus in Two Acts." *Small Axe* 12 (2): 1–14.

Hartman, Saidiya V. 2007. *Lose Your Mother: A Journey along the Atlantic Slave Route*. 1st ed. New York: Farrar, Straus and Giroux.

Heidegger, Martin. 2008. "Origin of the Work of Art." In *Martin Heidegger: The Basic Writings*, translated by David Farrell Krell, 143–212. HarperCollins.

Hollan, Douglas. 1995. "To the Afterworld and Back: Mourning and Dreams of the Dead among the Toraja." *Ethos (Berkeley, Calif.)* 23 (4): 424–36. https://doi.org/10.1525/eth.1995.23.4.02a00030.

Holstein, Kenneth, Erik Harpstead, Rebecca Gulotta, and Jodi Forlizzi. 2020. "Replay Enactments: Exploring Possible Futures through Historical Data." In *Proceedings of the 2020 ACM Designing Interactive Systems Conference*, 1607–18. DIS '20. New York, NY, USA: Association for Computing Machinery. https://doi.org/10.1145/3357236.3395427.

Hucal, Sarah. 2016. "What We Can Learn from Italy's Radical Design Movement." Curbed. December 21, 2016. https://archive.curbed.com/2016/12/21/14009200/florence-italy-radical-design-history.

Hunter, Tatum. 2021. "For Teens, Navigating the Mental Health Pitfalls of Instagram Is Part of Everyday Life." *Washington Post*, October 21, 2021. https://www.washingtonpost.com/techno logy/2021/10/21/teens-instagram-feed-mental-health/.

Irani, Lilly, Janet Vertesi, Paul Dourish, Kavita Philip, and Rebecca E. Grinter. 2010. "Postcolonial Computing: A Lens on Design and Development." In *Proceedings of the SIGCHI Conference on Human Factors in Computing Systems*, 1311–20. CHI '10. New York, NY, USA: Association for Computing Machinery. https://doi.org/10.1145/1753326.1753522.

Jackson, Steven J. 2014. "Rethinking Repair." In *Media Technologies*. The MIT Press. https://doi.org/10.7551/mitpress/9780262525374.003.0011.

Jasanoff, Sheila, and Sang-Hyun Kim, eds. 2015. *Dreamscapes of Modernity: Sociotechnical Imaginaries and the Fabrication of Power*. University of Chicago Press.

Jenkins, Tom. 2017. "Living Apart, Together: Cohousing as a Site for ICT Design." In *Proceedings of the 2017 Conference on Designing Interactive Systems*, 1039–51. DIS '17. New York, NY, USA: Association for Computing Machinery. https://doi.org/10.1145/3064663.3064751.

Jones, John Chris. 1970. *Design Methods: Seeds of Human Futures*. London, New York: John Wiley & Sons.

Jung, Heekyoung, Shaowen Bardzell, Eli Blevis, James Pierce, and Erik Stolterman. 2011. "How Deep Is Your Love: Deep Narratives of Ensoulment and Heirloom Status." *International Journal of Dsign* 5 (1). http://ijdesign.org/index.php/IJDesign/article/view/854/329.

Jungnickel, Kat. 2018. *Bikes and Bloomers: Victorian Women Inventors and Their Extraordinary Cycle Wear*. Goldsmiths Press. Cambridge: Goldsmiths, University London.

Jungnickel, Kat. 2021. "Speculatively Sewing Historic Clothing Patents." *Interactions* 28 (4): 15–17. https://doi.org/10.1145/3469863.

Kim, Raphael, Conor Linehan, and Larissa Pschetz. 2022. "Navigating Imaginaries of DNA-Based Digital Data Storage." In *Proceedings of the 2022 CHI Conference on Human Factors in Computing Systems*, 1–15. CHI '22. New York, NY, USA: Association for Computing Machinery. https://doi.org/10.1145/3491102.3501911.

Kimmerer, Robin Wall. 2015. *Braiding Sweetgrass: Indigenous Wisdom, Scientific Knowledge and the Teachings of Plants*. First Paperback edition. Minneapolis, Minn: Milkweed Editions.

Kirby, David. 2010. "The Future Is Now: Diegetic Prototypes and the Role of Popular Films in Generating Real-World Technological Development." *Social Studies of Science* 40 (1): 41–70. https://doi.org/10.1177/0306312709338325.

Kirk, David S., and Abigail Sellen. 2010. "On Human Remains: Values and Practice in the Home Archiving of Cherished Objects." *ACM Transactions on Computer-Human Interaction* 17 (3): 10:1–10:43. https://doi.org/10.1145/1806923.1806924.

Koskinen, Ilpo, John Zimmerman, Thomas Binder, Johan Redstrom, and Stephan Wensveen. 2011. *Design Research through Practice: From the Lab, Field, and Showroom*. Elsevier. http://books.google.ca/books?hl=en&lr=&id=TlsHSK9E3yoC&oi=fnd&pg=PP1&ots=rHu bY3cL0M&sig=o8TeLgGNrLvT9RkUApQMSXi9fos.

Kovach, Margaret. 2010. "Conversation Method in Indigenous Research." First Peoples Child & Family Review: An Interdisciplinary Journal Honouring the Voices, Perspectives, and Knowledges of First Peoples through Research, Critical Analyses, Stories, Standpoints and Media Reviews 5 (1): 40–48. https://doi.org/10.7202/1069060ar.

Kozubaev, Sandjar, Chris Elsden, Noura Howell, Marie Louise Juul Søndergaard, Nick Merrill, Britta Schulte, and Richmond Y. Wong. 2020. "Expanding Modes of Reflection in Design Futuring." In *Proceedings of the 2020 CHI Conference on Human Factors in Computing Systems*, 1–15. CHI '20. New York, NY, USA: Association for Computing Machinery. https://doi.org/10.1145/3313831.3376526.

Kuutti, Kari, and Liam J. Bannon. 2014. "The Turn to Practice in HCI: Towards a Research Agenda." In *Proceedings of the SIGCHI Conference on Human Factors in Computing Systems*, 3543–52. CHI '14. New York, NY, USA: Association for Computing Machinery. https://doi.org/10.1145/2556288.2557111.

Lang, Peter, and William Menking. 2003. *Superstudio: Life without Objects*. Milano: Skira.

Lewis, Jason Edward, Noelani Arista, Archer Pechawis, and Suzanne Kite. 2018. "Making Kin with the Machines." *Journal of Design and Science*, July. https://doi.org/10.21428/bfafd97b.

Lindley, Joseph, and Paul Coulton. 2015. "Back to the Future: 10 Years of Design Fiction." In *Proceedings of the 2015 British HCI Conference*, 210–11. British HCI '15. New York, NY, USA: Association for Computing Machinery. https://doi.org/10.1145/2783446.2783592.

Lindley, Siân E., Catherine C. Marshall, Richard Banks, Abigail Sellen, and Tim Regan. 2013. "Re-thinking the Web As a Personal Archive." In *Proceedings of the 22Nd International Conference on World Wide Web*, 749–60. WWW '13. New York, NY, USA: ACM. https://doi.org/10.1145/2488388.2488454.

Loonker, Mayank, Sophia Ppali, Rocio von Jungenfeld, Christos Efstratiou, and Alexandra Covaci. 2022. "'I Was Holding a Magic Box': Investigating the Effects of Private and Projected Displays in Outdoor Heritage Walks." In *Proceedings of the 2022 ACM Designing Interactive Systems Conference*, 1565–80. DIS '22. New York, NY, USA: Association for Computing Machinery. https://doi.org/10.1145/3532106.3533468.

Lunenfeld, Peter. 2003. *Design Research: Methods and Perspectives*. Edited by Brenda Laurel. Illustrated edition. Cambridge, Mass: The MIT Press.

Malpass, Matt. 2017. *Critical Design in Context: History, Theory, and Practices*. London ; New York: Bloomsbury Academic.

Mankoff, Jennifer C., Eli Blevis, Alan Borning, Batya Friedman, Susan R. Fussell, Jay Hasbrouck, Allison Woodruff, and Phoebe Sengers. 2007. "Environmental Sustainability and Interaction." In *CHI '07 Extended Abstracts on Human Factors in Computing Systems*, 2121–24. CHI EA '07. New York, NY, USA: ACM. https://doi.org/10.1145/1240866.1240963.

Massimi, Michael, William Odom, Richard Banks, and David Kirk. 2011. "Matters of Life and Death: Locating the End of Life in Lifespan-Oriented Hci Research." In *Proceedings of the SIGCHI Conference on Human Factors in Computing Systems*, 987–96. CHI '11. New York, NY, USA: Association for Computing Machinery. https://doi.org/10.1145/1978942.1979090.

Maturana, H. R., and F. J. Varela. 1991. *Autopoiesis and Cognition: The Realization of the Living*. Softcover reprint of the original 1st ed. 1980 edition. Springer/Sci-Tech/Trade.

McCarthy, John, and Peter Wright. 2004. *Technology as Experience*. Cambridge, Mass: MIT Press.

McCarthy, John, and Peter Wright. 2015. *Taking [A]Part: The Politics and Aesthetics of Participation in Experience-Centered Design*. Cambridge, Massachusetts: The MIT Press.

Mensonge, Kien. 2023. "Historically Informed HCI: Reflecting on Contemporary Technology through Anachronistic Fiction." *ACM Transactions on Computer-Human Interaction*, February. https://doi.org/10.1145/3517144.

Miyazaki, Hirokazu. 2004. *The Method of Hope: Anthropology, Philosophy, and Fijian Knowledge*. Stanford University Press.

Mollon, Max, and Annie Gentes. 2014. "The Rhetoric of Design for Debate: Triggering Conversation with an "Uncanny Enough" Artefact." In, 1049. https://shs.hal.science/halshs-01270043.

Morrison, Toni. 1987. *Beloved*. First Edition. Alfred A. Knopf.

Mozur, Paul. 2018. "A Genocide Incited on Facebook, With Posts From Myanmar's Military." *The New York Times*, October 15, 2018, sec. Technology. https://www.nytimes.com/2018/10/15/technology/myanmar-facebook-genocide.html.

Nemer, David. 2022. *Technology of the Oppressed: Inequity and the Digital Mundane in Favelas of Brazil*. MIT Press.

Odom, William, Richard Banks, David Kirk, Richard Harper, Siân Lindley, and Abigail Sellen. 2012. "Technology Heirlooms? Considerations for Passing down and Inheriting Digital Materials." In *Proceedings of the SIGCHI Conference on Human Factors in Computing Systems*, 337–46. CHI '12. New York, NY, USA: Association for Computing Machinery. https://doi.org/10.1145/2207676.2207723.

Ogbonnaya-Ogburu, Ihudiya Finda, Angela D.R. Smith, Alexandra To, and Kentaro Toyama. 2020. "Critical Race Theory for HCI." In *Proceedings of the 2020 CHI Conference on Human Factors in Computing Systems*, 1–16. CHI '20. New York, NY, USA: Association for Computing Machinery. https://doi.org/10.1145/3313831.3376392.

Oogjes, Doenja, and Ron Wakkary. 2022. "Weaving Stories: Toward Repertoires for Designing Things." In *Proceedings of the 2022 CHI Conference on Human Factors in Computing Systems (CHI '22)*. New York, NY, USA: Association for Computing Machinery. https://doi.org/10.1145/3491102.3501901.

Oulasvirta, Antti, and Kasper Hornbæk. 2016. "HCI Research as Problem-Solving." In *Proceedings of the 2016 CHI Conference on Human Factors in Computing Systems*, 4956–67. CHI '16. New York, NY, USA: Association for Computing Machinery. https://doi.org/10.1145/2858036.2858283.

Ozaki, Hiromi. 2010. "Menstruation Machine — Sputniko!" Sputniko! 2010. https://sputniko.com/Menstruation-Machine.

Pérez, Emma. 1999. *The Decolonial Imaginary: Writing Chicanas into History*. Bloomington: Indiana University Press.

Pierce, James, and Carl DiSalvo. 2017. "Dark Clouds, Io&#!+, and [Crystal Ball Emoji]: Projecting Network Anxieties with Alternative Design Metaphors." In *Proceedings of the 2017 Conference on Designing Interactive Systems*, 1383–93. DIS '17. New York, NY, USA: ACM. https://doi.org/10.1145/3064663.3064795.

Prado, Luiza. 2017. "Questioning the 'Critical' in Speculative & Critical Design." *A Parede* (blog). https://medium.com/a-parede/questioning-the-critical-in-speculative-critical-design-5a355cac2ca4.

Pschetz, Larissa. 2008. "{hi}stories: Supporting User Generated History." In *CHI '08 Extended Abstracts on Human Factors in Computing Systems*, 3693–98. CHI EA '08. New York, NY, USA: Association for Computing Machinery. https://doi.org/10.1145/1358628.1358915.

Puig de la Bellacasa, María. 2017. *Matters of Care: Speculative Ethics in More than Human Worlds*. 3rd ed. edition. Minneapolis: Univ Of Minnesota Press.

Redström, Johan. 2017. *Making Design Theory*. Cambridge, Massachusetts: The MIT Press.

Robles, Erica, and Mikael Wiberg. 2010. "Texturing the 'Material Turn' in Interaction Design." In *Proceedings of the Fourth International Conference on Tangible, Embedded, and Embodied Interaction*, 137–44. TEI '10. New York, NY, USA: Association for Computing Machinery. https://doi.org/10.1145/1709886.1709911.

Rosner, Daniela. 2023. "Fabulating Otherwise." *Interactions* 30 (2): 34. https://doi.org/10.1145/3582194.

Rosner, Daniela K. 2018. *Critical Fabulations: Reworking the Methods and Margins of Design*. MIT Press.

Rosner, Daniela K., Samantha Shorey, Brock R. Craft, and Helen Remick. 2018. "Making Core Memory: Design Inquiry into Gendered Legacies of Engineering and Craftwork." In *Proceedings of the 2018 CHI Conference on Human Factors in Computing Systems*, 1–13. CHI '18. New York, NY, USA: Association for Computing Machinery. https://doi.org/10.1145/3173574.3174105.

Sabie, Samar, Robert Soden, Steven Jackson, and Tapan Parikh. 2023. "Unmaking as Emancipation: Lessons and Reflections from Luddism." In *Proceedings of the 2023 CHI Conference on Human*

2017 Conference on Designing Interactive Systems, 503–14. DIS '17. New York, NY, USA: ACM. https://doi.org/10.1145/3064663.3064734.

Wakkary, Ron, Doenja Oogjes, Henry W. J. Lin, and Sabrina Hauser. 2018. "Philosophers Living with the Tilting Bowl." In *Proceedings of the 2018 CHI Conference on Human Factors in Computing Systems*, edited by Regan Mandryk, Mark Hancock, Mark Perry, and Anna Cox, 94:1–94:12. CHI '18. New York, NY, USA: Association for Computing Machinery. https://doi.org/10.1145/3173574.3173668.

Wallace, Jayne, Peter C. Wright, John McCarthy, David Philip Green, James Thomas, and Patrick Olivier. 2013. "A Design-Led Inquiry into Personhood in Dementia." In *Proceedings of the SIGCHI Conference on Human Factors in Computing Systems*, 2617–26. CHI '13. New York, NY, USA: Association for Computing Machinery. https://doi.org/10.1145/2470654.2481363.

Weiser, Mark. 1991. "The Computer for the 21st Century." *Scientific American* 265 (3): 94–104. https://doi.org/10.1038/scientificamerican0991-94.

Weiser, Mark. 1994. "Creating the Invisible Interface: (Invited Talk)." In *Proceedings of the 7th Annual ACM Symposium on User Interface Software and Technology*, 1. UIST '94. New York, NY, USA: Association for Computing Machinery. https://doi.org/10.1145/192426.192428.

Wensveen, Stehan and Ben Matthews. 2014. "Prototypes and Prototyping in Design Research." In *The Routledge Companion to Design Research* (pp. 262–276). Routledge.

White, Jordan, William Odom, Nico Brand, and Ce Zhong. 2023. "Memory Tracer & Memory Compass: Investigating Personal Location Histories as a Design Material for Everyday Reminiscence." In *Proceedings of the 2023 CHI Conference on Human Factors in Computing Systems*, 1–19. Hamburg Germany: ACM. https://doi.org/10.1145/3544548.3581426.

Williams, Rua Mae, Louanne Boyd, and Juan E. Gilbert. 2023. "Counterventions: A Reparative Reflection on Interventionist HCI." In *Proceedings of the 2023 CHI Conference on Human Factors in Computing Systems*, 1–11. CHI '23. New York, NY, USA: Association for Computing Machinery. https://doi.org/10.1145/3544548.3581480.

Winograd, Terry, and Fernando Flores. 1987. *Understanding Computers and Cognition: A New Foundation for Design*. Boston: Addison-Wesley.

Wu, Yiying, and Ilpo Koskinen. 2022. "Plant Hotels: Designing the Imaginary Foundations of Communities." *CoDesign* 18 (1): 32–47. https://doi.org/10.1080/15710882.2021.1991958.

Sterling, Bruce. 2009. "COVER STORY: Design Fiction." *Interactions* 16 (3): 20–24. https://doi.org/10.1145/1516016.1516021.

Suchman, Lucy A. 1987. *Plans and Situated Actions: The Problem of Human-Machine Communication.* New York, NY, USA: Cambridge University Press.

Tanenbaum, Theresa Jean, Karen Tanenbaum, and Ron Wakkary. 2012. "Steampunk as Design Fiction." In *Proceedings of the SIGCHI Conference on Human Factors in Computing Systems*, 1583–92. CHI '12. New York, NY, USA: Association for Computing Machinery. https://doi.org/10.1145/2207676.2208279.

Taylor, Alex S., Susan P. Wyche, and Joseph "Jofish" Kaye. 2008. "Pottering by Design." In *Proceedings of the 5th Nordic Conference on Human-Computer Interaction: Building Bridges*, 363–72. NordiCHI '08. New York, NY, USA: Association for Computing Machinery. https://doi.org/10.1145/1463160.1463200.

Taylor, Charles. 2004. *Modern Social Imaginaries.* Duke University Press.

Tharp, Bruce M., and Stephanie M. Tharp. 2019. *Discursive Design: Critical, Speculative, and Alternative Things.* Cambridge, MA: The MIT Press.

Thiry, Elizabeth, Siân Lindley, Richard Banks, and Tim Regan. 2013. "Authoring Personal Histories: Exploring the Timeline as a Framework for Meaning Making." In *Proceedings of the SIGCHI Conference on Human Factors in Computing Systems*, 1619–28. CHI '13. New York, NY, USA: Association for Computing Machinery. https://doi.org/10.1145/2470654.2466215.

Tillet, Salamishah. 2021. "Afrofuturist Room at the Met Redresses a Racial Trauma." *The New York Times*, November 17, 2021, sec. Arts. https://www.nytimes.com/2021/11/17/arts/design/metropolitan-museum-afrofuturist-seneca.html.

Tolmie, Peter, James Pycock, Tim Diggins, Allan MacLean, and Alain Karsenty. 2002. "Unremarkable Computing." In , 399–406. ACM. https://doi.org/10.1145/503376.503448.

Tonkinwise, Cameron. 2014. "How We Intend to Future: Review of Anthony Dunne and Fiona Raby, Speculative Everything: Design, Fiction, and Social Dreaming." *Design Philosophy Papers* 12 (2): 169–87.

Tsing, Anna Lowenhaupt. 2015. *The Mushroom at the End of the World: On the Possibility of Life in Capitalist Ruins.* Princeton: Princeton University Press.

Tucker, Holly. 2012. *Blood Work: A Tale Of Medicine And Murder In The Scientific Revolution.* 1st edition. New York: WW Norton.

Turner, Hannah, Laura Gibson, and Clara Gimenez-Delgado. 2021. "Participatory Design for the Anarchive: The Amagugu Ethu / Our Treasures Documentation Project." In *Designing Interactive Systems Conference 2021*, 1783–92. DIS '21. New York, NY, USA: Association for Computing Machinery. https://doi.org/10.1145/3461778.3462129.

Verbeek, Peter P. C. C. 2009. "Moralizing Technology: On the Morality of Technological Artifacts and Their Design." *Readings in the Philosophy of Technology.* https://research.utwente.nl/en/publications/moralizing-technology-on-the-morality-of-technological-artifacts.

Wakkary, Ron. 2021. *Things We Could Design: For More Than Human-Centered Worlds.* Cambridge, Massachusetts: The MIT Press.

Wakkary, Ron, William Odom, Sabrina Hauser, Garnet Hertz, and Henry Lin. 2015. "Material Speculation: Actual Artifacts for Critical Inquiry." In *Proceedings of The Fifth Decennial Aarhus Conference on Critical Alternatives*, 97–108. CA '15. Aarhus N: Aarhus University Press. https://doi.org/10.7146/aahcc.v1i1.21299.

Wakkary, Ron, Doenja Oogjes, and Armi Behzad. 2022. "Two Years or More of Co-Speculation: Polylogues of Philosophers, Designers, and a Tilting Bowl." *ACM Transactions on Computer-Human Interaction*, January. https://doi.org/10.1145/3514235.

Wakkary, Ron, Doenja Oogjes, Sabrina Hauser, Henry Lin, Cheng Cao, Leo Ma, and Tijs Duel. 2017. "Morse Things: A Design Inquiry into the Gap Between Things and Us." In *Proceedings of the*

Factors in Computing Systems, 1–15. CHI '23. New York, NY, USA: Association for Computing Machinery. https://doi.org/10.1145/3544548.3581412.

Sabie, Samar, Katherine W Song, Tapan Parikh, Steven Jackson, Eric Paulos, Kristina Lindstrom, Åsa Ståhl, Dina Sabie, Kristina Andersen, and Ron Wakkary. 2022. "Unmaking@CHI: Concretizing the Material and Epistemological Practices of Unmaking in HCI." In *Extended Abstracts of the 2022 CHI Conference on Human Factors in Computing Systems*, 1–6. CHI EA '22. New York, NY, USA: Association for Computing Machinery. https://doi.org/10.1145/3491101.3503721.

Schofield, Guy Peter. 2014. "Time Telescope: Engagement with Heritage through Participatory Design." In *Proceedings of the 2014 Conference on Designing Interactive Systems*, 117–20. DIS '14. New York, NY, USA: Association for Computing Machinery. https://doi.org/10.1145/2598510.2598517.

Seberger, John S. 2021. "Into the Archive of Ubiquitous Computing: The Data Perfect Tense and the Historicization of the Present." *Journal of Documentation* 78 (1): 18–37. https://doi.org/10.1108/JD-11-2020-0195.

Sedgwick, Eve Kosofsky. 1997. "Paranoid Reading and Reparative Reading; or, You're So Paranoid, You Probably Think This Introduction Is About You." In *Novel Gazing: Queer Readings in Fiction*, 1–38. New York: Duke University Press. https://read.dukeupress.edu/books/book/636/chapter/128566/Paranoid-Reading-and-Reparative-Reading-or-You-re.

Sengers, Phoebe, and Bill Gaver. 2006. "Staying Open to Interpretation: Engaging Multiple Meanings in Design and Evaluation." In *Proceedings of the 6th Conference on Designing Interactive Systems*, 99–108. DIS '06. New York, NY, USA: ACM. https://doi.org/10.1145/1142405.1142422.

Sharma, Sumita, Britta F. Schulte, Rocío Fatás, Noura Howell, Amy Twigger Holroyd, and Grace Eden. 2022. "Design Futuring for Love, Friendship, and Kinships: Five Perspectives on Intimacy." In *Extended Abstracts of the 2022 CHI Conference on Human Factors in Computing Systems*, 1–14. CHI EA '22. New York, NY, USA: Association for Computing Machinery. https://doi.org/10.1145/3491101.3516388.

Shin, Jo, Gabriela Aceves Sepúlveda, and William Odom. 2019. "'Collective Wisdom': Inquiring into Collective Homes as a Site for HCI Design." In *Proceedings of the 2019 CHI Conference on Human Factors in Computing Systems*, 1–14. CHI '19. New York, NY, USA: Association for Computing Machinery. https://doi.org/10.1145/3290605.3300546.

Soden, Robert, David Ribes, Maggie Jack, Will Sutherland, Vera Khovanskaya, Seyram Avle, Phoebe Sengers, and Susanne Bødker. 2019. "Fostering Historical Research in CSCW & HCI." In *Conference Companion Publication of the 2019 on Computer Supported Cooperative Work and Social Computing*, 517–22. CSCW '19. New York, NY, USA: Association for Computing Machinery. https://doi.org/10.1145/3311957.3359436.

Solnit, Rebecca. 2014. *Men Explain Things to Me*. Chicago, Illinois: Haymarket Books.

Søndergaard, Marie Louise Juul, Nadia Campo Woytuk, Noura Howell, Vasiliki Tsaknaki, Karey Helms, Tom Jenkins, and Pedro Sanches. 2023. "Fabulation as an Approach for Design Futuring." In *Proceedings of the 2023 ACM Designing Interactive Systems Conference*, 1693–1709. DIS '23. New York, NY, USA: Association for Computing Machinery. https://doi.org/10.1145/3563657.3596097.

Stappers, Pieter Jan. 2012. "Doing Design as a Part of Doing Research." In *Doing Design as a Part of Doing Research*, 81–91. Birkhäuser. https://doi.org/10.1007/978-3-7643-8472-2_6.

Statera, Gianni. 1979. "Student Politics in Italy: From Utopia to Terrorism." *Higher Education* 8 (6): 657–67. https://doi.org/10.1007/BF00215988.

Stengers, Isabelle. 2011. *Thinking with Whitehead: A Free and Wild Creation of Concepts*. Translated by Michael Chase. First English Language Edition. Cambridge, Mass: Harvard University Press.